Tributes
Volume 31

"Shut up," he explained.
Essays in Honour of Peter K. Schotch

Volume 22
Foundational Adventures. Essays in Honour of Harvey M. Friedman
Neil Tennant, ed.

Volume 23
Infinity, Computability, and Metamathematics. Festschrift celebrating the 60th birthdays of Peter Koepke and Philip Welch
Stefan Geschke, Benedikt Löwe and Philipp Schlicht, eds.

Volume 24
Modestly Radical or Radically Modest. Festschrift for Jean Paul Van Bendegem on the Occasion of his 60th Birthday
Patrick Allo and Bart Van Kerkhove, eds.

Volume 25
The Facts Matter. Essays on Logic and Cognition in Honour of Rineke Verbrugge
Sujata Ghosh and Jakub Szymanik, eds.

Volume 26
Learning and Inferring. Festschrift for Alejandro C. Frery on the Occasion of his 55th Birrthday
Bruno Lopes and Talita Perciano, eds.

Volume 27
Why is this a Proof? Festschrift for Luiz Carlos Pereira
Edward Hermann Haeusler, Wagner de Campos Sanz and Bruno Lopes, eds.

Volume 28
Conceptual Clarifications. Tributes to Patrick Suppes (1922-2014)
Jean-Yves Béziau, Décio Krause and Jonas R. Becker Arenhart, eds.

Volume 29
Computational Models of Rationality. Essays Dedicated to Gabriele Kern-Isberner on the Occasion of her 60th Birthday
Christoph Beierle, Gerhard Brewka and Matthias Thimm, eds.

Volume 30
Liber Amicorum Alberti. A Tribute to Albert Visser
Jan van Eijck, Rosalie Iemhoff and Joost J. Joosten, eds.

Volume 31
"Shut up," he explained. Essays in Honour of Peter K. Schotch
Gillman Payette, ed.

Tributes Series Editor
Dov Gabbay   dov.gabbay@kcl.ac.uk

"Shut up," he explained.
Essays in Honour of Peter K. Schotch

edited by
Gillman Payette

© Individual authors and College Publications 2016. All rights reserved.

ISBN 978-1-84890-187-2

College Publications
Scientific Director: Dov Gabbay
Managing Director: Jane Spurr

http://www.collegepublications.co.uk

Cover design by Laraine Welch

Printed by Lightning Source, Milton Keynes, UK

All rights reserved. No part of this publication may be reproduced, stored in a retrieval system or transmitted in any form, or by any means, electronic, mechanical, photocopying, recording or otherwise without prior permission, in writing, from the publisher.

Contents

Introduction
GILLMAN PAYETTE 1

1 Portrait of the Artist as a Logician
PETER K. SCHOTCH 5
  1.1 Introduction 5
  1.2 Apology 6
  1.3 I am born 6
  1.4 Graduate Study in Philosophy at Waterloo during the Boom 8
  1.5 What I "learned" at Waterloo 8
  1.6 My Arrival At Dalhousie 9
  1.7 Ray Jennings and Others 10
  1.8 My Students 14
  1.9 My List of Projects 16
  1.10 The way I work 20
  1.11 My Prose Style 21
  1.12 What I Think is Important 22
  1.13 Bibliography 24

I Papers in Peter's Honour 25

2 C.I. Lewis and Entailment
EDWIN MARES 27
  2.1 Introduction 27
  2.2 The Early Development of Lewis's Theory of Entailment 28
  2.3 Global and Local Conventions 29
  2.4 The Logics 30
  2.5 Justifying Logical Principles 31
  2.6 Transitivity of Strict Implication 33
  2.7 The Consistency Postulate and S1 36
  2.8 What Lewis Entailment is Not 37
  2.9 Concluding Remarks 37
  2.10 Bibliography 38

3 The Logical Project
Letitia Meynell and Richmond Campbell — 41
- 3.1 Introduction — 41
- 3.2 What is Logic? — 43
- 3.3 The Logical Project — 51
- 3.4 What is Logic? Revision — 55
- 3.5 Conclusion — 58
- 3.6 Bibliography — 59

4 Identity, Haecceity, and the Godzilla Problem
Kent A. Peacock and Andrew Tedder — 63
- 4.1 Introduction: An Embarrassing Problem — 63
- 4.2 A Short History of Arguments for =I — 64
- 4.3 Blocking the Godzilla Inference — 66
- 4.4 Haecceities, or the Lack Thereof — 68
- 4.5 The Null Object — 72
- 4.6 Summing Up — 75
- 4.7 Bibliography — 78

5 Alice Munro's "Wild Swans"
Steven Burns — 81
- 5.1 A Philosophical Thesis — 81
- 5.2 Introduction to "Wild Swans" — 82
- 5.3 Introduction to Lloyd's Evolutionary Biology — 83
- 5.4 Crime, Coercion, and Consent — 85
- 5.5 Conclusion of Lloyd's Project — 89
- 5.6 Conclusions About the Story — 92
- 5.7 Bibliography — 95

6 Socrates the Janus-Faced: The *Logos* of the Phaedrus
Leon McQuaid — 97
- 6.1 The Logic of '*Logos*' — 97
- 6.2 *Logos*: Inference, Memory and Discourse — 99
- 6.3 A Country Setting — 102
- 6.4 The Opening Gambit: Lysias' Speech — 103
- 6.5 Phaedrus rebuffed: Socrates' First Speech — 107
- 6.6 Socrates' Second Speech: The Janus-Faced — 111
- 6.7 *Logos* and *Alogos* — 114
- 6.8 Conclusion — 115
- 6.9 Bibliography — 116

II Papers Related to Schotch's Work — 117

7 Scotch on Quine, with Help From Carnap and Prior
M. J. Cresswell — 119
- 7.1 Truth Functions and Necessity — 119
- 7.2 Prior on Truth — 124
- 7.3 Conclusion — 131
- 7.4 Bibliography — 134

8 Globalization Makes Inconsistency Unrecognizable
John Woods — 137
- 8.1 Ex Falso Quodlibet — 137
- 8.2 Damage Control — 141
- 8.3 Unrecognizability — 151
- 8.4 How Come? — 154

9 From Jónsson and Tarski to Schotch and Jennings
Alasdair Urquhart — 167
- 9.1 Introduction — 167
- 9.2 Boolean Algebras with Operators — 168
- 9.3 Diagonalizing Multimodal Operators — 170
- 9.4 Representation and Duality for Objects — 172
- 9.5 Duality for Maps — 174
- 9.6 Composition of Multimodal Operators — 176
- 9.7 The Finite Model Property and Decidability — 177
- 9.8 Bibliography — 179

10 From Consequence Relations to Reasoning Strategies
M. Bryson Brown — 181
- 10.1 Introduction — 181
- 10.2 Forcing and Variations — 181
- 10.3 Chunk and Permeate — 184
- 10.4 Level-preservation in C and P — 185
- 10.5 Applications — 187
- 10.6 Bibliography — 188

11 Type Raising: Schotch on the Prisoner's Dilemma and Deontic Logic
Gillman Payette — 191
- 11.1 Introduction — 191
- 11.2 Type Raising — 192
- 11.3 Schotch's Take on the Prisoner's Dilemma — 194
- 11.4 Type Raising in Deontic Logic — 202
- 11.5 Conclusion — 216
- 11.6 Bibliography — 217

Introduction

I first met Peter though his reputation. In 2000 I was a first-year commerce undergraduate at Dalhousie. I quickly saw that commerce was not for me, so I switched to math. Something that mirrors Peter's own beginnings at university. Because of the switch, my schedule opened up. I signed up for a logic course to sate a curiosity I picked up in high school which is traceable, ultimately, to Peter.[1] Peter didn't teach that course; it was another near miss.

In the math department, I expressed a curiosity for logic, and Peter was known as 'The Logician'. He was the, by all accounts eccentric, person I should work with. I am not sure what that says about what the mathematicians thought of me. I recall an older student commenting that 'that logician wears a baret', people that use math are not usually that stylish—maybe Nicolas Bourbaki was. Finally, in 2002-3, the chance came to take the advanced logic course. Most of us in the course were honours math students. We thought we could breeze through the course. We were wrong. It wasn't that we didn't have the chops; we were not accustom to using math in the way that Peter was using it. The math comes second, so the philosophy comes first.

The other thing that comes first for Peter: his students. I am the product of that, career-wise. I owe him a debt that I can never really repay. Not only bringing me in to philosophy by funding my Master's project, but also letting me stay with him and Carra during my Killam Postdoctoral fellowship at Dalhousie. It was intellectually stimulating and terribly fun. I think it was largely responsible for me getting the Banting fellowship; it gave me the time to work up that project and spend a lot of time with my family in Calgary. But enough about me.

I think I should say something about this collection of papers in honour of Peter. First, I chose the title since it is one of Peter's best lines. So much so that the undergrads at Dalhousie had it printed on hoodies for their society. (It is my second favourite hoodie.) Using that infamous line as a title for this book seems appropriate since it also displays Peter's sense of humor (he agreed that it was a good title). I think Peter has always thought a bit of irreverence is a good thing. The line also offers a way lighten the mood before responding to a particularly good criticism. Philosophy is serious enough, it doesn't need our help.

---

[1] It is actually quite a surprising coincidence. When I was in high school I was acquainted with Jorgen Jensen's daughter (Jorgen was a student of Peter, as you will read in his autobiography). Jorgen had logic books from his time as Peter's student, and those piqued my curiosity. I ended up in logic, in a roundabout way, because of Peter's influence. Nova Scotia is a small place.

Second, I have broken up the papers into two parts. The first part are papers which honour Peter. They deal with topics that Peter finds interesting. I know Peter enjoys the intersection between philosophy and literature, and really enjoys teaching it because it shows how philosophical movements can remain relevant (particularly with existentialism). So while he doesn't publish on that subject, he appreciates work in that vein. Ed Mares' paper deals with a topic that has been of active interest to Peter over the past several years: C. I. Lewis' work on logic and why Lewis thought S3 was the right system of strict implication. Peter has expressed deep respect for Ed's scholarship on that topic, so I am glad that some of it has been included here.

Kent Peacock and Andrew Tedder's paper which questions some of the basic rules of classical first-order logic: existential introduction. Peter loves to question the basic rules. They call it the 'Godzilla problem', but I am tempted to call it 'the contemporary problem of existential import'—I have a flair for making things sound dull. The paper presents a worthwhile discussion of the problem and maybe what to do about it. Rich Campbell and Letitia Meynell's paper offers up an interesting take on the grounding of logic in the naturalistic project after Quine's "Epistemology Naturalized". I think Peter would be more sympathetic to their project than they think. Peter is really a 'reflective equilibrium' kind of logician: logical consequences find their justification by balancing theory and intuition. Logic isn't handed down by God or a Platonic heaven.

In the second part of the book are papers which touch on Schotch's work. Bryson Brown and Alasdair Urquhart present some of the work within the "preservationist" school of logic. Brown offers an overview of paraconsistent logic in the preservationist school, and the augmentations of it—Chunk and Permeate—which model reasoning with inconsistency in the history of science. Urquhart generalizes the Tarski and Jónsson "algebraic" semantics for modal logic to n-ary modal logic. N-ary modal logic is one of the unique contributions of Schotch and Jennings. Max Cresswell offers an interpretation and critique of some of Peter's recent work on the meaning of logical connectives, and puts that discussion in the context of early/mid 20th century discussions of the meaning of logical connectives.

John Woods explores the upshot of paraconsistency generally. And makes some comments on/criticisms of the preservationist approach to paraconsistency. He defends his own moderate way to deal with inconsistency without the heavy lifting required by preservationist solutions nor simply denying the validity of *ex falso*. In my essay I explore some thoughts of Peter's on social choice/The Prisoner's Dilemma, as well as some of his related thoughts on deontic logic. I organize this around the central topic of comparing sets of objects by comparing the elements of those sets: a topic Peter calls 'Type Rasing'.

Finally, I have asked Peter to include a bit of an intellectual autobiography. I have always loved Peter's stories, and I thought that this would be a way to get him to include some in this book. I have spent a lot of time with Peter, and I have learned much about how to do philosophy and logic from that. My thinking is that this would be a way for others to spend some time with Peter as well.

Now comes time for *my* thanks. I must say thank you to all those who submitted essays, and express my appreciation for the patience everyone has shown me as I have put this together over three years. As well, I should thank the Social Sciences and

Humanities Research Council of Canada for the Banting Fellowship which funded my research while I edited this collection. Thank you!

Gillman Payette
Calgary, Alberta, August 2016

Chapter 1

Portrait of the Artist as a Logician

Peter K. Schotch[1]

1.1 Introduction

My academic career has not gone the way I thought it would. Not at all. As a young academic, I thought that after a few years of active research, I would run out of steam. I recognized even then that while I was very intellectually ambitious, I had plenty of examples of equally ambitious people who started early, like I did, but were just going through the motions by the time they hit their middle years. I'm not talking about the notorious "dead wood" which seems to many members of the North American public to afflict universities in general. I'm talking about people who may well be publishing up a storm, and doing exemplary teaching, committee work, and service to the community. However, much of their research consists of adding bells and whistles to the work of their younger selves or supervising students who are adding the bells and whistles. Students need thesis topics after all. I was ready for that to happen to me. It didn't.

"Why is that?", I hear you cry—I'm not at all sure. Partly I keep on because I am stubborn, and partly it's because I am seldom satisfied by my own work. But mostly I am inclined to think that I still have something that others have lost altogether or have retained only a version of it withered by time and the usual slings and arrows of academic and ordinary life.

I find it hard to describe this mysterious fount of motivation without using somewhat questionable prose (and not for the first time, as I discuss in the sequel). It's as if I can sense something, something significant and important. It may appear like a tiny cloud on the horizon no larger than a man's hand. But I sense it none the less and it calls to me. So if I reduce my efforts now, it will decrease my chances of making the discovery.

Above, I used the rather unkind phrase "run out of steam" to describe those who no longer feel the pull that I feel, or not to the same extent as I. But this kind of description is more derogatory than is warranted. It is possible, and even likely, that the academics I am talking about no longer feel the pull because they have *made* the important discovery

---

[1] Dalhousie University, peter.schotch@dal.ca

the existence of which was signaling them on its own private channel. Once one has dug up the buried treasure, it seems a bit silly to keep on digging. Of course one mustn't entirely discount greed (worst construction) or obsessive compulsion (a somewhat better construction).

Early in my career I was taken to task by Alex Rosenberg who claimed to be distressed at the amount of work I was doing compared to the amount of "deliverables" I produced.[2] My reply to him was that while some of the stuff I was writing might well have been publishable, I maintain that there is a distinction between publishable work and work which ought to be published. He shook his head at the depth of my delusion, but I reckon that History has proved me right. The profession is currently reeling under the weight of published material which ought not to have been published. With the recent appearance of online journals which can easily and cheaply put out "issues" of journals which contain thousands of pages has made matters far worse. The motivation for this innovation is evidently financial for the journal owners and job-related issues for those who provide the content and for whom "publish or perish" is not some quaint dictum of the last century.

1.2 Apology

It will not have escaped the keen attention of the reader that much if not most of this essay is devoted to acknowledging my intellectual debts. I have tried to do this as completely as I can, but I am certain that this is not nearly complete enough. To those of you who deserve mention but fail to receive it, please accept my profound apologies. My once fabled memory has by now become merely storied.

1.3 I am born

I was born in the middle of the 20th Century (in 1946, to be more precise). This is important since it has certain implications as regards my logical training. I entered the undergraduate program at the University of Waterloo in 1965-66. when Waterloo had around 3500 students in all faculties, but even then it had acquired an enviable reputation in engineering and informatics. It had also, by means of some political and financial slight-of-hand, managed to put together a very impressive department of philosophy.

Remember these were the years that put the boom in baby-boomers. In short, there was plenty of money for those who knew how to get hold of it, and as I intimated, the Waterloo department, and indeed the university as a whole, was quite deft in that regard. This also was important to my education as I shall explain below.

I began my undergraduate career in political science—an accident occasioned by my advisor at registration who assumed that I had come to him because he was a political scientist. In fact I had come to him because he had the shortest line. That should have raised a red flag, but I was a callow and unsophisticated babe in the woods.

---

[2]This is unfair to Alex. He has never, to my knowledge, used the word "deliverables."

1.3. I AM BORN

My budless career in PoliSci came to a grinding halt in the first lecture of a class entitled 'Canadian Public Administration,' in my opinion the most boring class offered anywhere. One of my electives was a philosophy class, in fact Logic 100 or similar. I was finding the class interesting and engaging so, on the spot, I changed my major to philosophy with help from the departmental administrative assistant—a wonderful helpful and sympathetic woman who actually ran the department, the imaginings of various chairpeople notwithstanding.

My first textbook in logic was the one by Barker. It was well-known at the time and may be still for all I know. I didn't like the book much, it had lots of stuff but I got the feeling that it was dumbed-down more than a bit. In my second course, the textbook was the infamous one by Copi. I hated that book. Fortunately, for me, there was another section of the second logic class that used Church's famous *Introduction*. The class was taught by Don Roberts, known mainly for his work on Peirce's logic (existential graphs), but he himself had taken a class with Church (using the latter's book). I think that Roberts was more comfortable as an historian of logic rather than as a logician. He often seemed less that entirely sure of himself.

Part of the problem may have been that Roberts had great difficulty in getting up in the morning. In an attempt to cure this he signed on to teach his logic class in the 8:00-10:00 time slot in the morning of Mondays and Wednesdays. Alas this did nothing to cure Roberts, if anything it made things worse. After missing a few classes because of oversleeping, he was forced to stay awake all of Sunday and Tuesday nights. So his students were faced with a somewhat strung-out and bleary Don Roberts telling us all about how to do logic à la Church. I liked the class because I liked the book rather than any stellar pedagogy. Roberts used what I understand was Church's method of evaluation. Each of us was required to write a textbook for the class. Many of the students hated this, but I enjoyed it very much. That fact, as much as any other led to my specializing in formal logic.[3]

I next took the graduate class in logic (also available to advanced undergraduates—one demonstrated such advancement by applying to take the class). It is fair to say that this class changed my life and is responsible for many of my logical attitudes to this day. The instructor was J. Sayer Minas, my most important early influence. I don't think that Jay was a brilliant teacher, although there were certainly flashes of that, but it was more a matter of his attitude to the material. His class was not of the familiar "Here's what the book says—any questions?" sort which is more likely to produce zombies than students. In fact, the book was Rosser's *Logic for Mathematicians*, which we were expected to read but hardly received a mention in class. Assuming we knew the received view, Jay began to ask us if we thought it was correct and to suggest ways in which it might be lacking. It is exactly *that* mind set which has stayed with me, while much of the rest of Jay's views on the nature of logic have since fallen away. Somebody once said to me:

---

[3] I had a different experience from Steve Martin as recorded in his book *Born Standing Up*. Martin having done well in his introductory symbolic logic class decided to enter the advanced symbolic logic class at Berkeley. His earlier class had not prepared him properly for a full frontal exposure to leading edge formal logic and he decided not to become a philosopher. I am certain that the world of entertainment and letters' gain was, for us, a bitter loss.

> You can always tell one of Minas' students—where other people are working on topological categorical recursion theory, a student of Minas is more likely to be working on conjunction.

Mind you, that should not be taken to imply that we students of Jay didn't know any of the tricky mathematical stuff. The opposite would be true. While at Waterloo I took a very influential course on Lattice Theory (that was its title at least) from Denis Higgs. I also took the usual classes in linear algebra, group theory etc. I had a course in number theory from Laslo Kalmar, and had my appetite for category theory whetted by several different people, first among whom was Higgs. I actually got to meet F.W. (Bill) Lawvere and hear an early version of his paper "Quantifiers and sheaves."

In fact for the first year I was enrolled in a rather ambitious joint doctoral program in philosophy and pure mathematics. Alas after many administrative and other disputes, that program was eliminated leaving those of us enrolled to pick only one of the two. I chose philosophy, and I think everybody else chose mathematics. My reason for the choice was that it was the only way I was going to get to work with Minas.

1.4 Graduate Study in Philosophy at Waterloo during the Boom

It must be very hard for the graduate students of today to understand what it was like to be one of the doctoral candidates in philosophy at Waterloo in the late 1960's and early 1970's. It was because the Ontario universities were funded by an enrollment driven formula. According to this formula graduate students were worth more than undergraduates and doctoral students more than masters students. This explains why a relatively small and (then) undistinguished university had 84 philosophy doctoral candidates on the books while having only 14 undergraduate majors.

Suffice it to say that there was money for everything and everybody. Four of us graduate students went to the Chair (don't quite recall who that was at the time) and complained that there wasn't enough logic to satisfy all those interested in the subject. I'm not actually certain that we didn't exaggerate just a touch. In response to the question of what we suggested be done, we asked for a number of short (two or three lectures) courses by visiting luminaries. These included Alonzo Church, Stephen Kleene, and Hilary Putnam among others. Next week it was all set.

1.5 What I "learned" at Waterloo

Perhaps it would be better to say "what I was taught and what I learned at Waterloo." In the main I left the doctoral program at Waterloo believing that I was an analytic philosopher and also a logician. That was quite an achievement since many of the people in the department at that time didn't believe that anyone *could* be both of those things. Certainly other attempts, such as Peter Strawson's *Introduction to Logic* failed to attract much attention from either camp.

Apart from the (major) influence of Jay Minas' logic teaching, the logic I learned at Waterloo was mostly what was called "mathematical" logic. This included the usual mix of foundations of mathematics stuff along with some recursion theory and the like. Looking back, I now see that I never got anything more than a glance at proof-theory. This may or may not explain my subsequent development as I explore below.

1.6 My Arrival At Dalhousie

I arrived at Dalhousie University where I have spent most of my subsequent career in the philosophy department. The job market was then (in the early 1970's) nearly as barren as it has become in the early 21st century which raises the question "How is it that you, Peter Schotch, were able to get what was the only tenure stream job in Canada that year?" In truth, by luck as much as anything. It turned out that the year I was hired, a group of members of the department became interested in the sorites problem and had read a highly technical and rather puzzling paper on vagueness and infinite valued logic by a mathematician named Goguen.[4]

I had listed among my areas of competence "many-valued logic", and luckily for me, I also indicated an interest in epistemology and Descartes. This meant that I wasn't a total logic nerd so the department gave me an interview. During the interview they discovered that I too had read the Goguen paper along with a few others and had an interest in infinitely many valued logic.

All that was not nearly enough to carry the day. The extra bit, the one that made the real difference was that the two other candidates for whatever reason did not interview well. This is by no means to say that they were not as good or as well-trained as I—I very much doubt that. I had unfortunately used up the good luck, and they had to make do without. That's what I mean by luck.

Having written that just now, it occurs to me to wonder how I managed to carry the whole thing off. I had come from a community in the Waterloo philosophy department where, honesty compels me to say, that I was not particularly well regarded except by three or four (out of 24 full time tenured professors). I was not even widely thought to be well regarded. So how did I parlay something of a reputation for arrogance, for not finishing what I started (about which a great deal more, below), and for not being concerned with "real philosophy" into an actual job?

I'm not at all sure I can answer that question in any very satisfactory way. One mustn't forget that this young man we are discussing was a creature who lived at a time not that far from the middle of the last century. This is a general problem with autobiography. I was that person, it is quite true, but I am no longer. The changes are significant and were I somehow to meet this person today, I doubt I would recognize him. This severely limits my status as an authority on the thoughts and feelings and intellectual ambitions of the younger version of me. On the contrary, my reports of those things should be viewed with suspicion since it is far too easy to inject bits of my current thoughts into the reports without even being aware that I am doing that.

---

[4] I think this was his one and only foray into the area.

1.7 Ray Jennings and Others

I arrived at Dalhousie thinking that the semantic approach to logic was the primary one. During my first year Ray Jennings visited the department. This visit had been arranged through the good offices of Steven Burns who had been in the doctoral program at the University of London (Bedford College) at the same time as Ray. Since Ray and I were both interested in logic and since my department was trying to manage this visit on the cheap, Ray stayed with me for five days. We had a wonderful time (or perhaps it would be more correct to say that I had a wonderful time and Ray seemed not to mind).

Just imagine, this was the very first time in my life that I had another logician to talk to for an extended period. I realize that this seems an odd thing to say. Of course I had years of talking to logicians, but these were brief conversations and I often had the feeling that we were all too much on our guard for anything very important to happen as the result of the talk.

With Ray things were different. I never got the feeling with him that he was holding anything back lest I get some kind of advantage over him and I certainly never felt like holding back whatever was my latest thinking on various projects. This was a strong foundation for our future friendship and collaboration.

It is important to keep in mind that Ray and I came from different traditions in the study of modal logic—which is what I think first attracted us to one another. I had been pursuing the kind of logic which followed C.I. Lewis and H.B. Smith (or so I thought at the time) largely through the study of the kind of finite Boolean models pioneered by Huntington. Ray on the other hand was fresh from a year visiting Victoria University in Wellington New Zealand. There he had the advantage of studying with Max Cresswell, George Hughes, Robb Goldblatt and a number of other leading lights. In fact I felt quite put in the shade.

For one thing Ray was an absolute master of the Henkin completeness proof where my own grasp of the technique was somewhat attenuated. I learned a tremendous amount from Ray at that first meeting. It also helped a lot that we were, both of us, deep in the laws of thought tradition, according to which a logic was identified with its set of theorems. An additional link was the fact that the two of us were used to coming at the various issues of modal logic from the semantic perspective, although rather different ones. That meeting was to set the foundation for a lot of our future work together.

I won't record here how, in the following summer Ray and I stumbled, with the help of a comment from Barbara Partee, from more-or-less pure modal logic to a study of what became the beginnings a new[5] approach to paraconsistent reasoning. I won't tell that story here because I have already told it in some detail in the introductory chapter of *On Preserving*.

---

[5]Honesty, again, compels me to admit that my assertion of novelty might be challenged by the friends of David Lewis, who always claimed that he had the idea first. His account was first published as "Logic for equivocators" in *Nous*, several years after some of our publications. Of course for all I know he may have first had the idea in 1972 or even earlier, no matter when he first published it. In my humble opinion however there really isn't any full-blooded anticipation of our approach in the Lewis paper his footnote about Ray and me notwithstanding.

1.7. RAY JENNINGS AND OTHERS

Ray and I did two paper-reading tours of, respectively, Europe and Australasia. The European tour was where we both met Johan and Krister for the first time and, as I note below, presumed on the basis of an acquaintance best described as scraped, to impose ourselves on our gracious hosts.

We had a great time with the van Benthems, with most of our logical discussion held in the evening. The topics were wide-ranging and the discussion usually went on until midnight or so. The next morning without fail, Johan would present us with a set of notes concerning what had transpired the evening before, but these were far from being minutes. This is because Johan would usually supply an original proof or two to answer questions which had been raised. This was, to put it mildly, a humbling experience.

Next we stayed with Krister Segerberg and his wife in Turku (southern Finland). The hospitality shown us was breathtaking. We took the very welcome chance to present some of our stuff (the modal part of it) at the Finnish university where we were delighted to make the acquaintance of Risto Hilpinen. We were also treated to an evening at the Swedish club where Ingemar Pörn read a paper which he had recently written. He began reading in Swedish and then switched to English, at a request from Krister on behalf of his barbaric guests who spoke no Swedish. It was impressive, to say the least.

After Scandinavia we returned to England where Ray and his family were staying in a small village near Lewes. Here we spent a lot of time at the local pub, and very nice it was too, mostly discussing what lessons we thought we had learned on our trip, and trying to rough out some sort of completeness proof for our version of paraconsistent logic. I think we managed some baby steps but the proof in question was still several months away, and a really convincing proof, a couple of years.

We took a break from this rather frustrating job to visit both Oxford and Cambridge (in two separate trips). In Oxford we spent an afternoon with Kit Fine who was borrowing Dana Scott's office just before going on to Salzburg. We started with lunch and then went on from there. Ray had been worrying at a question in modal logic which he mentioned to Kit. Crunch crunch grind grind— You could almost see the wheels going round. Then after a couple of minutes, Kit had the answer. This was a bit scary, though very impressive of course.

Then we moved on to talk about our conviction that a certain kind of modal logic lay at the heart of paraconsistent logic. Kit thought about that for a bit and then pronounced it interesting. We didn't heave a great sigh of relief, but we certainly felt a kind of weight lifting. It's all very well to be a voice in the wilderness, crying, but it helps to have a few people listening.

In Cambridge we visited with Tim Smiley at Clare College. He was able to put us up in the college and we had most enjoyable conversation about the beginnings of the (then) recent explosion in research in modal logic. By coincidence, while we were walking we came upon Casimir Lewy. Tim Smiley very kindly introduced Ray and me as visiting authorities in modal logic, a title which we hardly deserved. We had a nice chat with one the legendary pioneers of the question which is central to our notion of paraconsistent logic. But there was better to come. After going back to Tim's rooms we sat around and explained our notion of a hierarchy of levels of inconsistency. He was silent for a bit and then came the accolade: "Damn, I wish I had thought of that!"

No amount of praise can beat that, something that I have kept in mind every since.

Our trip to Australia in the early 1980's was under the auspices of Richard Routley (who later changed his last name to Sylvan—it's a long story which I won't tell here). Alas Richard himself was not there when we arrived since, ironically, he was visiting Canada at the time.

In New Zealand and Australia we once again took advantage of several people's hospitality. Ross Bryant, Michael McRobbie and Max Cresswell in particular. Looking back on the tour, I now think that the paper reading part was the least successful. That was because, for the most part, the people we were talking to had their own research programs from which it would have taken more a single talk by Ray and me to deflect them. An attempt to give a talk to the philosophers at Otego University in Dunedin was a particularly notable disaster.

On the other hand, the conversations we had were what a more civilized century would have called "improving." And, I should hurry to say we had a *lot* of conversations. As a result of this I came to a better understanding of the program of relevance logic(s) as well as a much better appreciation of the work of Charles Hamblin (which would stand me in good stead in my own work on the logic of rules) as well as Frank Vlach.

So now comes the question that has been preying upon all of your minds for many years. Why did the collaboration between Ray and me cease? Was a woman involved or perhaps a drinking problem? Did one of us try to take credit for work done by the other?[6] None of these things happened. Ray and I remain the best of friends to this very day but our interests began to diverge.

I think that this divergence began with the day I first read Dana Scott's version of the Lindenbaum lemma for his (Scott's) sort of sequent logic. To say that I was impressed by his proof would be a considerable understatement. I thought then, and I think now that the proof (in spite of an annoying typo) is one of the most elegant and beautiful I have ever experienced. On top of this, Scott's proof seemed to open up a new and fruitful area of knowledge. And almost that quickly I was through the looking glass. Semantics would never again be quite as important in my thinking about logic as it had been. I was no longer happy to identify a logic with its set of theorems and began to identify a logic with its set of valid (in some generic sense of that term) inferences.

This led within a relatively short time to Ray and I publishing a paper called "Inference and necessity." The title was mine. In that paper I dragged Ray into asserting with me that there is a natural connection between inference relations and necessity operators. Mind you we really didn't say much more than that in the paper and, after years of thinking about it I doubt I can say very much more that would be intelligible.

But while I was on fire with my newfound conversion to the inferential model of logic and my swerve in the direction of proof-theory, Ray was not even a bit singed around the edges. It wasn't long before Ray and I had a conversation in which Ray told

---

[6] Actually the issue of who did what remains shrouded in mystery in lots of our early work. More than once Ray has attributed to me, work that I have attributed to him. This is not a case of "After you Alphonse," it is rather a case that neither of us can recall the precise fact of the matter. It hasn't helped that in our joint writing, we each got the knack of writing the way we thought the other person wrote. In fact, of course neither of us *actually* wrote as the other would have written but in a kind of blend of the two styles. So textual analysis is unlikely to help with the question of the true authorship of this or that paragraph.

1.7. RAY JENNINGS AND OTHERS

me that his own interests had diverged from mine (a very kind construction to put upon the events) and that it might be better if we ceased writing together. Of course I was a bit stricken by this since I very much enjoyed writing with Ray, but I could certainly see his point. And as I said earlier, our parting was amicable and we remain fast friends.

Apart from Ray, there were lots of people who influenced my logical development. Max Cresswell was an early and continuing influence and a great source of encouragement. Hughes Leblanc also went out of his way, in some case quite far out of his way to encourage me and Ray also. I owe them both a great debt which can only be properly repaid by encouraging others. This I have tried to do, and have kept on. But I cannot rid myself of the feeling that I have not done nearly as good a job as either Max or Hughes.

Others who must be mentioned as having had a noticeable influence on my development as a logician, include Bas van Fraassen, Kit Fine, Steve Thomason, Brian Chellas, Alasdair Urquhart, and many others. I would like to highlight in particular Johan van Benthem, and Krister Segerberg who, with their families, extended hospitality far beyond the call of duty. I only wish that they come and visit me in Nova Scotia for a week or (better) two. The autumn is particularly beautiful here although July and August are usually reliable too. Best of all, come at the same time—we have lots of room.

Thus far I have left the impression that nobody at Dalhousie has had any influence on my work at all. This is quite wrong. In my own department Bob Martin encouraged me to the point of writing a paper with me. This was quite important at that stage of my career. But the main influence was that of David Braybrooke. From him I learned, in the first place, that the Harvard "debating" model of doing philosophy was not the only way of doing things and in fact was inferior to a more collegial approach. This was extremely important to my subsequent development.

Apart from that valuable insight, David began to invite me to work on various papers and projects with him. His main interest was the way in which philosophy could play a role in the social sciences, and I was especially eager to assist him in these projects. In the end David and I received a substantial grant[7] from the SSHRC (social sciences and humanities research council of Canada) for the purpose of completing a project on the logic of rules as applied to what David called 'large scale history.' This grant enabled us to hire Bryson Brown as our research assistant, thus rescuing him from a (no doubt) high paying job outside of philosophy. Thus began the project that eventuated in *Logic on the Track of Social Change* (Oxford). This was an interesting project in lots of ways. First it aimed at contributing to something outside of logic, indeed outside of philosophy properly speaking. Of course there was plenty of philosophical content but it clear to all of us that it was going to be a hard sell.

David was by no means a logician but he was an expert in social and political philosophy. He was also convinced that logic had a role to play in the study of rules and had already made a start using some early work of von Wright. Bryson and I convinced him that it would be better to redo this right from the start. For us that meant trying to figure out, at least roughly, what one wanted the logic of rules to do, and only starting to frame the logic after that. I think this is the correct way to do it, but also that we were

---

[7]The story of precisely how we obtained this grant is curious and interesting but too long for this essay.

a bit too brisk in many spots. In any event it would be something of a stretch to say that the book was well received. The logicians thought (in general) that we had left too much undone, and the non-logicians couldn't see the point of 'all that weird logic stuff'.

I would say my major contribution to the logic was to insist that we take actions seriously. This means, for me, that we make reference to actions in the object language and that, semantically, such reference is irreducible in the sense that we don't take actions to reduce in some way to propositions.[8]

I also received a good deal of encouragement and help from the department of mathematics at Dalhousie. Shortly after I arrived I met Robert Paré who was, since Lawvere had left the year before, the principle category theorist. I sat in on Bob's class on topoi, and learned not as much as I should have but, as someone once said, enough to be dangerous. Within a year or two there began a weekly seminar called the Atlantic Category group (@cat), Which was far ranging enough that I was invited to give talks several times over the years. I found the participants to have considerably more intellectual good will than most of the philosophy groups to whom I gave talks. I now believe that my current views on the algebra of logic have been heavily influenced by my association with @cat.

It would be quite wrong of me to leave this section without recording my gratitude to the Society for Exact Philosophy. I missed very few of the meetings of the Society during my "formative" years. The influence of all the papers I heard and discussions in which I participated had an effect on me which it would be difficult to overestimate. It is a pleasure for me to record my thanks and to acknowledge my debt to the SEP and all who sail in her.

1.8 My Students

It is both easy and difficult to recognize the contribution of my students. The easy part is to acknowledge my debt to certain outstanding students, although even in that case I might forget certain people who should have been mentioned. To all those, my sincerest apologies.

The outstanding students fall into two categories, those who wanted to participate in some active way in one or more of my projects, and those whose main contribution was some telling criticism or even more telling insight. I fear it's the latter group who shall receive short shrift. But more about them later.

The first students I recall who made a strong impression on me were Jurgen Jensen and Sharon Logan. For some reason they were interested in many valued logic and helped me to generalize the Henkin-style completeness proof for the Łukasiewicz 3 valued logic, to all the Łukasiewicz finitely many valued logics. Not a grand contribution and most who understand the original paper on 3-valued logic, could undertake the generalization. But none of them did and so my students got a publication out of it.

---

[8]For example we don't take all actions to be the bringing about of the truth of some or other proposition. This is not to say, of course, that there are no actions of that type, but only that such actions do not exhaust the class of all actions.

1.8. MY STUDENTS

During this time I also became interested in the completeness issue for Łukasiewicz infinite valued logic. I had been through the proof by Rose and Rosser but that was only for week completeness. I thought for a while that using the Henkin method might secure the stronger result. I agree that this was naive and unscholarly of me, but don't forget that there was at this time only a remark by Rosser in a rather obscure publication, that one of his students had shown that weak was as strong as it gets. After agonizing over this with my students for a week or so, it finally occurred to me that we had accumulated enough material to prove that the infinite valued case was non-compact. Since compactness follows at once from the Henkin proof, there can be no such proof for infinite valued logic. The end.

To the best of my knowledge, this result hasn't been published until recently.[9] On the day I sent the result to Steve Thomason who was fiddling around with many-valued logic to ask whether he thought it was publishable. He said no, except as part of a larger study of three valued logic. In the end, that is exactly how it came to be (finally) published in *Excursions*. Neither Jurgen nor Sharon went on in philosophy, which didn't surprise me. I think they both found the subject interesting but a bit too placid for them.

My next standout was Blaine d'Entremont. Blaine came a bit later to philosophy having majored in mathematics at nearby St. Mary's University. Blaine was my student during the end of the time when Ray and I were working together, so perhaps the double dose of logic helped to give him a boost. He was of great help to us in developing a sequent logic approach to the paraconsistent inference relation we were already calling "forcing". In addition he developed some very useful ideas about compactness in his MA thesis.

It was during this time that a kind of pool of students formed in the philosophy department. This group, which included Blaine, had as it's logical core: Paul Thompson, Ann Levey, Judy Pelham, and Mary MacLeod along with others, no doubt, whom I have forgotten.

It wasn't that all these students wanted to become logicians; most of them were more interested in other subjects. But all of them had some interest in logic and in talking about logic, asking questions, and all the other things that can make teaching the very opposite of a burden.

I look back on this as a kind of golden age for logic at Dalhousie. I suspect that there are very few who can look back to two such ages, but I am one of the lucky ones. There was also a twenty-first century age; which also led to quite a bit of logical research. The person most responsible is Gillman Payette.

Gillman came into my life as a student in the second logic course, at a time when my work with David Braybrooke was drawing to a close. It never quite finished drawing, but that is more often the case with me than not. I was initially impressed with Gillman's energy and enthusiasm as well as his burning intellectual ambition. He was also talented but that usually goes with the other qualities. In him I recognized something very close to my younger self, but, in spite of that we got along very well and continue to do so.

I had been waiting quite a long time for some of my fellow countrymen or, at least their students, to take up some of the many open problems which remained in our

---

[9] Schotch, Peter: *Excursions in Philosophical Logic*, Chapter 4, Amazon 2010

paraconsistent logic program which, by then, had come to be called "preservationism."[10] One must have a care when mentioning projects to Gillman. I mentioned a few of these in connection with his MA thesis which he was shortly to begin.

He fairly tore into these problems in a manner I had never before witnessed in either student or colleague. Not only did his work throw light on the problems and questions I had raised, it led to the discovery of new and important issues which had slipped between the cracks of the original work. I also noticed, it was impossible not to, that the more he worked, the better logician he became. With lots of nagging and criticism from me his prose and presentation skills also improved by leaps and bounds.

My only problem with Gillman was that he made me lazy. Seeing that something or other ought to be proved, rather than do it myself, I would assign it to Gillman. I know that pattern is quite common between teachers and students but it still strikes some people as unfair since it is too easy at that point to slide into appropriating the work of the student or, nearly as bad, assigning oneself a crucial role in having achieved the result. In my own defense I will say only that I would far rather my student prove something than I do and that I always wear my non-slip shoes. In the end, I feel privileged to have had Gillman as a student and collaborator and I recognize that most academics are not as lucky as I in having a student like that at least once in their career.

1.9 My List of Projects

Throughout most of my career, I have been interested in modal logic. A lot of my working and thinking over the years has been about foundations both from the technical end of things and the philosophical one too. I've also been a student of the history of the subject though, for the most part, my scholarship has a been a bit patchy to say the charitable thing. I like to think that this has improved as time goes by, but then I sometimes like to think that Deans are fine and upstanding people.

On the technical side of things, I have been a bit dismayed by the turn towards mathematization, as I suspect several of you have. I know very well that applying mathematics to logic has a grand and shining tradition, but it seems to me quite important that we keep in mind that we are mathematizing something, viz. logic, that is not mathematics, no matter what the mathematicians may say. There is, in other words, a line here though it is often unclear sometimes where it lies.

I'll have more to say about this below. For now, back to modal logic. The first problem one encounters as a writer on foundational issues is that most people think there are no such issues, that the entire matter concerning foundations was settled in the 1960's and 70's. Kripke appeared upon the scene and suddenly we all knew what necessity and possibility meant. To quote David Lewis on the subject:[11]

> Necessity is simply a restricted universal quantifier.

---

[10] And this is another annoying example in which I attribute this coinage to Ray, and he to me—I'm starting to believe in the existence of a person my students call SchotchAndJennings, while Ray's refer to JenningsAndSchotch.

[11] *Counterfactuals*

1.9. MY LIST OF PROJECTS

In this he echoes a remark by Bertrand Russell in his *Lectures on Logical Atomism*, except that Russell says nothing about any restriction. Most famously Russell claims that the whole field of modal logic is mostly to be laid at the doorsteps of those who have not received the notion of a propositional function. This is very close to his criticism of the Scottish logician Hugh MacColl's way of representing modal functions.

Nor is this a minority opinion among the great mass of philosophers without a sound working knowledge of the issues. The nutshell version goes:

> Before Kripke came on the scene philosophers had a very dim understanding of the real meaning of the modals. But afterwards, all the mystery and fog has cleared up and we know, except perhaps for some very arcane technical stuff, what necessity and possibility are all about.

I don't think I'm being unfair here, but this view offends the true history of 20th century modal logic as well as a couple of other things. In explaining that I shall reveal the roots of some of my projects.

1.9.1 The Truth of the Matter

What we call the semantics of classical propositional logic came to us out of 19th century algebra. In a similar manner the semantics of modal propositional algebra came to us out of 20th century algebra. In particular Tarski and his first North American graduate student, Bjarni Jónsson, published part one of their "Boolean Algebra with operators" in 1951.[12]

In the first historical footnote in his 1963 paper[13] Kripke refers to the earlier work as a "surprising anticipation". I'm inclined to think that remark a bit cheeky. I doubt very much that Tarski was surprised.

As for things being a muddle before the advent of so-called possible worlds semantics, I think this is based on hasty or, in some cases, non-existent scholarship. In an attempt to do something about this I have been trying to illuminate the work of C.I. Lewis and H.B. Smith in their pre-war work. Far too often people have never heard of Smith and look at Lewis only through the lens of Quine.[14]

1.9.2 Yet Another Problem

Lots of consumers of modal logic or, at least possible worlds semantics, think some wrong things. As pointed out above lots of people, some of them quite clever, think that Lewis is right about necessity simply being a universal quantifier. They think, in other words that modal logic is another way of doing ordinary predicate logic. This is, of course, false. We know thanks to the work done by Johan van Benthem, Rob Goldblatt,

---

[12](Jónsson and Tarski, 1951)

[13](Kripke, 1963)

[14]As the outcome of this effort on my part there are two essays in my *Essays in Philosophical Logic*[15] entitled respectively "Modal logic in the early twentieth century: C.I. Lewis" and "Modal logic in the early 20th Century: H.B. Smith".

and Steve Thomason that modal propositional logic is as expressive as the full second order theory of one binary relation. Perhaps modal logic is another way of doing set theory, but not of doing first order logic.

Since these results, which arrived to perplex us in the early 1970's, require a heavily technical argument to establish, the ordinary practitioners generally deal with this problem by not talking about it.

And last, we have among us many who believe (1) that possible world semantics à la Kripke is correct and (2) That S5 is the correct logic of (logical) necessity since it is obvious that necessity simply means (alethic) truth in all possible worlds. I have argued[16] that K is actually the nearest we can get to the logic of logical necessity, and that S5 is not determined by the "truth in all possible worlds" condition. The latter is a relatively easy consequence of the fact that modal classes of frames are closed under disjoint unions.

1.9.3 At Last, Some Projects

Having lived through the 1960's I, like so many others, had high hopes for applications of the new logic. And, like so many others, I have seen my hopes deflate. Of course I'm rather picky and there have been others who thought comfort and succor, if not actual philosophical progress, can be derived from modal considerations. I think that most of these people have got it wrong.

Consider the several attempts to apply modal logical lore to the free will issue. From a neutral perspective it's all a dog's breakfast. One person derives the basis of free will from a form of the K axiom (called by these folk the "main modal principle") while another (or perhaps the same person) derives the basis for determinism from that very same principle. All I can think to say in this situation is "That's not helpful!"

Many people thought that Fred Dretske was onto something important in a way that Hintikka wasn't, when "Epistemic operators" appeared (Dretske, 1970). But I don't think it's difficult to see that there really aren't glad tidings for either epistemology or moral philosophy.

I argue in *Excursions* that epistemic logic cannot be normal, thus preventing the use of the logical machinery that most people were hoping to use. I use another argument, even more pointed perhaps, to show that what most people understand by the term "deontic logic" is impossible.

So, if the attempted applications of logic in general and modal logic in particular fail, is there any hope that some other attempt might not fail, or at least not fail so obviously. I think that there is. In two separate projects I try my hand at establishing a new formalism for epistemic logic[17] and showing that formal methods in moral philosophy need not collapse under their own weight.[18]

Nor have I given up trying to make sense of "ordinary" modal semantics. During the start-up, people were prone to wonder what to make of the binary relation given

---

[16] In *Excursions* and elsewhere
[17] In *Excursions*.
[18] In *Ethics as a Formal Science*, forthcoming.

1.9. MY LIST OF PROJECTS

as part of the modal frame. The pioneers were quick to claim this as a feature rather than a bug. Since no assumption is made about this relation the user is free to impose whatever interpretation interests them more. Perhaps, as Kripke himself suggests, we might produce some interpretation which would fit deontic logic. Alas, this won't work. And the same goes for almost all other suggested interpretations[19] save only one. When we interpret the mysterious modal relation as the temporal before-after relation we break through onto a broad plateau of new and interesting results. We are made to think over a lot of the philosophy of time that we thought we new.

Why is this? Why is it that the promise of a cornucopia of new applications to philosophy soon shriveled to a single new thing—temporal logic. In the arguments I have mentioned previously it is because for most applications, normal modal logic commits one to too many suspect principles at the very outset.

Many years ago, it occurred to me, as I am sure it did to some of you, that perhaps we had things backward. Perhaps instead of temporal logic being an application (and the most successful one at that) of modal logic, modal logic descends, in some fashion, from it. If something along these lines could be worked out, I speculated, we would be able to define the modal relation from the before-after relation, thus removing a mystery and source of anxiety.

I worked on this for many years and the study led me to study many-dimensional logic. It often seemed I was getting close to something worth-while but close wasn't enough. Finally I decided to show some of this work to Gilman and that was enough to get me over the hump. With Gilman's help I saw that what I was struggling with was a not-unknown kind of system called hybrid logic. Seeing that opened virtually all of the technical doors and indeed eventuated in some interesting temporal considerations including a new approach to Diodorus' so-called master argument. Much of this work was published in an article entitled "Worlds and times." (Schotch and Payette, 2011).

But the question with which I started really needs more work. Can the modal relation be defined? I think so and I think the definition can be done inside the system of Worlds and times. Some of this later work is reported in my essay "The way back into the ground of modal logic."[20]

Of course I remain concerned with the foundations of logic. In the 19th Century logic came out of algebra. This is not to say that logic *is* algebra as so many mathematicians would have it. I have come around to thinking that logic does indeed arise from certain algebraic considerations. The difference is that I now think that the algebra in question is not Boolean but rather categorical (a 20th Century invention) or perhaps homotopy type theory (21st Century). Of course these latter two are not incompatible—it's more a matter of emphasis.

I have yet to publish any of this algebraic work, which should not come as a surprise. Or, if it does, read the next section.

---

[19] Apart from helpful analogies like "seeing", used to good effect in {citepIML
[20] In my *Essays in Philosophical Logic*

1.10 The way I work

Given what I have already written, it will surprise nobody that I tend to work slowly. Sometimes, I'll confess, I enter a period of low energy (to which I usually refer as hibernation). But the whole matter is more complicated than mere idleness can explain.

Sometimes, especially early in one's career, it makes sense to work and publish as fast as you can. There is an article which could help in a tenure or promotion hearing, if only it can see the dark of print in time. There is a conference that would be good to attend, for academic or other reasons (it's not for me to judge). If a certain project could be made into a publication then one or more students would have it on their CV. More often than not several of these motivations are combined to urge one into print.

As opposed to this, there is the matter of the important distinction between the work that is publishable and the work that ought to be published, which I mentioned earlier.

All of these things are germane, but the reason I work (or at least publish) so slowly has a different explanation. I must say that it makes me a bit uneasy to reveal this reason. When Johan van Benthem was looking at my CV with a view to writing a letter recommending me for something or other (I can't remember now what it was but I'm sure that Johan's letter helped a lot), he wrote to ask me why there were these substantial gaps in my list of publications. Rather than tell him the whole truth which, as I said, is a bit embarrassing, I told him that the gaps represented times when I was working on books, or other more substantial projects. Of course I'm sure Johan remained puzzled since he was able to work on books and several other projects at the same time. But, perhaps he thought, not all of us are able to multi-task to the same extent.

Of course, there was some truth to my reply to Johan, but it was not the whole truth.

Here's, the thing: There you are. There is your manuscript already in the form of a pdf file let's say. The submission process is usually trivial, a matter of uploading the file to your chosen journal's website. So why not just go ahead and do it?[21]

There is a spectrum here. At the pathological end are people like Henry Sheffer who was convinced that anything he sent away would be appropriated by others who would then become famous as a result of his work. I'm not like Sheffer. There are perfectionists, people who cannot look at their own writing without finding something about which to complain. I take Friedrich Waismann to have been such a person. He was said to have withdrawn the same manuscript many times over so that he could fix this or that mistake, as he saw it. I'm not really like Waismann either.

Sometimes people are worried that they are about to publish something containing embarrassing errors. I do worry about that, though probably not as often as I should. It's happened to me twice in my career but I soothe myself by thinking that at least they were interesting mistakes.

At the end of the day, what stops me from rushing into print, is something else. I nearly always feel that what I have written is not actually wrong, it is incomplete. I am bedeviled with the thought that not only have I missed something, but that what I missed is important. I don't actually feel this way about absolutely everything I have published—that would make of me a Sheffer.

---

[21] As the Nike shoe maxim has it.

It happens enough though, to explain various gaps in my publication record. Not perhaps in any full-blooded sense of 'explain,' but this is the best I can do.

1.11 My Prose Style

It was towards the end of my undergraduate career that I realized that philosophers *had* a prose style. I knew very well that there were such things as prose styles and had spent time studying Jane Austin, Joseph Conrad, and the other usual suspects in literature classes. But it never dawned on me that Philosophers might have such a thing as well until I took a course on Bertrand Russell. After that I began to notice that other writers in philosophy had a recognizable and in some cases, admirable "voice." Why did this discovery take so long? It's because most contemporary philosophical writing is badly, and in some cases very badly, written. Now, at the beginning of the twenty-first century, philosophers routinely publish things that would have been rejected out of hand a century earlier, not on grounds of content, but rather on grounds of form.

I've been thinking about and trying to work on my own prose; trying to find my most effective "voice," since I realized that without such thought and work, my prose would limp along as flaccidly as that of so many others. It seemed important to me then and it still does to this very day, to steam against this tide. In this as in so many other things Ray Jennings was a significant spur. From the start I admired Ray's prose as well as his presentation skills. I think I have learned a lot from him, especially during the time when we wrote together.

This is not to say that Ray and I write in the same way. We would both deny that.[22] But this is certainly true: I write the way I do more because of my acquaintance with Ray than for any other reason.

But what exactly is that way? It's far from clear that I'm the best person to say, but I am duty-bound to try. I prefer clean prose that does not vary too much from the way I speak—when I'm speaking about philosophy. Some people (you know who you are) find that a bit too informal. There is what one might call the European tradition, in which formality is often *de rigeur*, and some of those folk are not entirely comfortable with my prose.

But isn't that absolutely what one should expect in the world of letters? Some people don't like Hemingway's prose. I would be uncomfortable trying to write in the style of Hemingway and, no doubt, would produce only an inferior example of his sort of prose. But this is not to say that I deplore Hemingway. In just the same way I have not shrunk from collaborating with people whose writing was far more formal than mine.

For several years I worked with David Braybrooke on the logic of rules. I think we managed the collaboration quite well in spite of the fact that his prose (as I once said to him) could have landed him a job as a screenwriter for 18th Century Fox.

I would prefer to describe my prose as light-hearted. It reflects the fact, when it's any good, that I enjoy writing. I'm sure there are others who would prefer the description lightweight. I wasn't aware until I was reading a referee's comments on a piece of

---

[22]In fact Ray has told me more that once that I write more like Graham Priest than anyone else. I take that as a complement.

my work that there was such a word as 'jokey'. It turns out that there is indeed such a word and it suffers considerably in comparison to 'humorous' or 'funny.' Well, *nil disputandum* and all that. I don't like brussels sprouts while perhaps my referee loves them to distraction. But I can't say that without being disingenuous. You see, I believe that we can meaningfully dispute all sorts of esthetics and in fact do so every day.

But I want to know if jokeyness in general is displeasing or just my particular brand of it. Of course there are subjects which, in certain contexts, cannot be treated in a (consciously) humorous way, without one's taste being called into question. I'm fairly sure though, that paraconsistent logic is not one of those subjects. I'm not sure of the best way to handle such comments except to say that I personally know that some of my readers like the way I write because they have said as much to me. Although I should admit that these sayings all took place in spoken conversation. In print, I suspect that the closest my prose comes to approval is "accessible." I'll take it.

But I mustn't leave this topic without noting that a more serious criticism has been leveled at my prose which does not flow from a lack of appreciation of my 'jokeyness'. In fact the very opposite is true. Here is a quote from a review of *On Preserving*:[23]

> Against these criticisms Scotch (*sic*) and Jennings, either together or with co-authors, defend preservationism in a witty and sometimes mocking style. That makes their expositions a funny read, but it sometimes papers over a lack of argument on their part.

Here I have at least made the transition from jokey to witty (unless Ray's contribution made that difference). [spoiler alert: jokey] But now only imagine! I have the power to cloud the minds of my readers so that, be they ever so vigilant, they miss noticing that I have stopped arguing. That's pretty unsettling if not terrifying. I have sworn to use this power only for good.

1.12 What I Think is Important

In this section I'm going to preach a bit. I suspect that I'm preaching to the choir, but even choristers need some straight talk once in a while.

As I hinted at above, logicians, are often regarded by their colleagues as borderline philosophers at best, and non-philosophers who are just flim-flaming people with a lot of techno-babble that nobody really understands, at worst. This is largely a North American thing as far as my experience goes, but there are plenty of us in North America.

Perhaps this is simply a reflection of the thorough-going and rather vicious anti-intellectualism which has afflicted both the United States and, to a lesser extent, Canada. The grand tradition has given us climate change deniers, anti-vactionationists, moon landing conspiracy theorists and ultimately, Donald Trump. I simply don't know, but I *do* know that the effect is real.

---

[23] Bremer,Manuel,Philosophy in Review XXX (2010), no. 6,430-431

1.12. WHAT I THINK IS IMPORTANT

Logicians act variously in response. Some rise above it secure in their own value and get to the point where they don't even notice the odd sleight. Some retreat from the *real* philosophers becoming rather hermit-like in their own departments. Some become embittered which eventually shows up in both their teaching and their writing.

I think it important that we don't give in to this systemic bias against us. I am not uttering a call to the barricades I hasten to add—kicking against the pricks can only make things worse for us. I advocate consciousness raising instead. This is why I spent that time working with David Braybrooke, and why most of the talks I give in my department's colloquium series are about the role logic can play in other areas of philosophy. Sometimes, and perhaps even more importantly in the matter of inculcating sympathetic awareness, I talk about the role other areas of philosophy can and do contribute to philosophical logic.

Without efforts like this, our colleagues see less and less of our relevance to their own thought. We need to exert ourselves to see this doesn't happen or at the very least that it doesn't happen in the dark.

How did it work out for me? I can report only partial success. For some of my career I have felt like a poor relation in my department and, at times, it has certainly got me down. At other times I felt more visible and I found that made a noticeable difference to my teaching.

I think it important that we spend more time working on our curricula. Too many of us, including me embarrassingly often, are bogged down in old stuff that gets harder and harder to motivate, unless one takes a mainly historical approach. As a matter of fact, far too few of us are much concerned with the history of logic and would do well to improve our acquaintance with it. I have often found it a source of stimulation, and not just in my teaching.

Having raised the issue of teaching, here is something we really ought to take more pains to correct: Since the start of my teaching career I have noticed something odd. The grades in the introductory logic class typically have the pattern that the female students, on average, do better than the male students. This, even though the very top grade goes to a male more often than a female. I should also say that typically there are a similar number of each gender.

In the upper level logic classes the grades tend to level out more but the distribution of males to females in these classes tips heavily in the direction of the males. Why is this? I don't have the answer but I wouldn't be surprised if feminists weren't surprised. The situation of female students in the 'hard' sciences seems of a piece with my observation.

At this stage figuring out where to point the finger of blame is not going to help here. What will help? That's difficult to say. What I personally have done is to identify those female students with aptitude in logic and encouraging them to enter the upper level classes. It's my impression, unsupported by statistics, that I have trained more female logicians than anybody else in Canada. But lest you think me a braggart, you should notice that the bar is not set very high.

1.13 Bibliography

Dretske, F. (1970). Epistemic operators. *Journal of Philosophy*, 1–22.

Jónsson, B. and A. Tarski (1951). Boolean algebras with operators. I. *Amer. J. Math. 73*, 891–939.

Kripke, S. A. (1963). Semantic analysis of modal logic i, normal propositional calculi. *Zeitschrift für mathematische logik und Grundlagen der Mathematik 9*, 67–96.

Schotch, P. K. and G. Payette (2011). Worlds and times. *Synthese*, 295–315.

Part I

Papers in Peter's Honour

Chapter 2
C.I. Lewis and Entailment

Edwin Mares[1]

2.1 Introduction

This paper is dedicated to Peter Schotch. During the year I spent at Dalhousie in the early 1990s, Peter was going through an S1 phase. During that year, and afterwards, I spent many hours discussing possible semantics for S1 with Peter. This was also my first introduction to the philosophy of logic of C.I. Lewis. I found Peter's thoughts on this topic, as on all topics in logic and philosophy, full of interesting ideas, only some of which I really understood at the time. More recently, Peter and I have engaged in a discussion about Lewis's notion of entailment. This paper presents my side of that conversation.

It is common now to treat C.I. Lewis's logics S1, S2, and S3 as being of merely historical interest. The semantics for those logics are inelegant, especially when contrasted with those of the standard normal modal logics, such as K, T, S4, and S5. To criticise Lewis's logics for their lack of an illuminating model theory is reasonable. It clear that Lewis himself wants a coherent theory of meaning. He makes several remarks in his earlier work about the meaning and much of his later work is taken up with that topic. But Lewis is not only concerned with the semantics of his logics. Lewis is a pragmatist. He thinks of logic as a tool. The central use of this tool is in theory construction. And theory construction, as Lewis understands it, requires a logic like S1, S2, or S3. S4 and S5 are not well suited to this task. This is why Lewis says:

> Those interested in the merely mathematical properties of such systems of symbolic logic tend to prefer more comprehensive and less 'strict' systems, such as S5 and Material Implication. The interests of logical study would probably be best served by an exactly opposite tendency. (Lewis and Langford, 1951, p 502)

---

[1] Department of Philosophy, Victoria University of Wellington, edwin.mares@vuw.ac.nz

The purpose of this paper is to examine the role of logic in Lewis's epistemology and then to use this role to justify Lewis's choice of logical systems. At the end of the paper, I use this justification to defend Lewis's logics against a common criticism.

2.2 The Early Development of Lewis's Theory of Entailment

In his earliest papers, Lewis did not construct the logic of strict implication as a tool for theory construction. Rather he wanted a more intuitive formalisation of the notion of inference and proof. In a 1911 letter to his undergraduate logic teacher, Josiah Royce, Lewis says:

> I am quite convinced now of the possibility of modifying the calculus of propositions so as to bring its meaning of implication into accord with that of ordinary inference and proof. I worked up a preliminary paper, and would trouble you with it if I had a legible copy. Since it only costs postage, I have sent it to "Mind."[2]

In order to justify the logic of strict implication, in his pre-1920 writings Lewis's relies on intuitions about the ordinary notion of proof and a contrast with the main opposing view – the logic of *Principia Mathematica*. In a 1920, we see an important change in Lewis's view of logic. He rejects his earlier descriptive and psychologistic view of logic Lewis (1921).

What happens in 1920 is that Lewis turns sharply towards pragmatism and abandons his earlier Kantian idealism. He develops a philosophy he calls *conceptual pragmatism*. One of the elements of this new philosophy is the pragmatic treatment of necessity. A statement is necessary if it is *a priori*. It is a priori, moreover, if it is stipulated in a way that makes it impervious to empirical data. For example, the truths of arithmetic are necessary in this sense for Lewis. The adoption of the pragmatic theory makes Lewis a conventionalist in a sense. I will return to his conventionalism and its relationship to his logic is section 2.3.

A second element of conceptual pragmatism that is important for my current topic is the idea that we construct our view of the world in terms of the *systems* that we accept. A system is what logicians now usually call a "theory", except that whereas a theory is a set of formulas or sentences, a system is an extra-linguistic structure. A systems is a collection of states of affairs (which Lewis's calls "facts"), that is closed under provable implication and conjunction:

> In the meaning which we shall assign, any set of facts constitutes a system if it meets the following requirements:
>
> 1. If A is a fact of a given system – call it Σ – and A is inconsistent or incompatible with B, then B is not in Σ.
> 2. If C is in Σ, and C requires or implies D, then D is also in Σ.

---
[2]Letter from Lewis to Royce, October 15, 1911. Harvard University Archives, Royce Papers, Box 122, folder 44.

3. If E and F are facts in Σ, then Σ contains also their joint-fact EF.
(Lewis, 1923a, p 385)

In the notes for his 1923 lectures on logic at Columbia University, Lewis makes clear that by "requires or implies" here, he means strict implication.[3] There is nothing in the definition of a system that precludes its being empty, but I will only consider populated systems – systems that contain at least one state of affairs.

An example of a system that Lewis uses quite often is that of an applied geometry. A geometry is developed as a "conceptual system in the abstract". When it is applied by an agent as a description of physical space, the considerations that affect the choice are all pragmatic (Lewis, 1929, p 298). Lewis's move away from Kantianism can be seen in his treatment of a priori systems, like that of a geometry. He thinks of geometry in a modern axiomatic and conventionalist manner. A geometry is understood determined by a set of axioms – all of the statements of the geometry can be derived by pure logic from these axioms. The geometry then gives us empricial criteria for determining what is a straight line, what is a triangle, and so on (ibid. pp 298f).

In the 1920s and early 1930s, Lewis treats his modal logics to a large extent as theories of the closure of systems. It is my contention that we can see in this for logic justification for the logical principles that Lewis accepts and rejects.

2.3 Global and Local Conventions

Before I examine the logics, I need to present some more background. From 1920, Lewis adopts what he calls a "pragmatic theory of the a priori" and of necessity Lewis (1923b, 1929). There are two aspects to this theory: (1) what is a priori or necessary is conventional; (2) what is a priori or necessary is independent of experience.

Lewis's conventionalism, at least from 1920 until the mid-1930s, is an extreme sort. He maintains that all our beliefs that we think are necessary are in fact chosen by us and imposed upon experience. We could adopt radically different a priori beliefs. We could think that space has a different geometry, we could use very different categories and classifications, and we could divide the world into individuals in very different ways. Lewis even thinks that we can choose a different formal logic. The choice between logics is based entirely on pragmatic considerations Lewis (1932); Lewis and Langford (1951).

A belief is independent of experience if it is consistent to hold that it is true regardless of what experiences one has. For Lewis any belief held to be necessary must be independent of experience. Even physical laws are independent of experience:

> the experience which fails to conform to the law is repudiated as non-veridical. (Lewis, 1929, p 261)

If an agent has an experience that does not seem to obey what she accepts as a law of nature, she has the choice of thinking of that experience as being illusory, hallucinatory, or otherwise no-veridical.

---

[3] Lewis, "Columbia Lectures on Symbolic Logic 1923-1927", Stanford University Library, Department of Special Collections, M174, Box 20, folder 20-1.

In one sense, Lewis does not distinguish between different sorts of necessity. All necessity is conventional and all necessity is independent of experience. But Lewis does distinguish between more and less general necessary truths (Lewis, 1929, p 246). What exactly he means by this might seem obscure, but I suggest that it can be understood in the following manner. Logical necessity, for Lewis, is the most general sort of necessity. In adopting a system of logic as a logic, one commits herself to understanding all of her other systems as closed under that logic. We can generalise this idea. There are systems that are subordinate to other systems. Consider a model, in the scientific sense, of a particular physical theory. That model contains particular parameters and initial conditions that are not in the theory itself, but is closed under the laws of the theory. In this way, we can think of an agent's systems as making up a hierarchy. Those laws, such as the laws of logic, under which every system in this hierarchy is closed can be thought of as global laws and those under which only some systems are closed as more local laws.

2.4 The Logics

The purpose of this paper is to develop a Lewisian justification of the logics S1, S2, and S3. Lewis accepted S3 as his logic of strict implication from the *Survey of Symbolic Logic* Lewis (1918) until some time between 1927 and 1930. His formulation in the survey is flawed. Emil Post showed that the logic trivialises. Lewis presented a corrected form first in Lewis (1920). Lewis still accepted S3 as his system of logic through his 1927 lectures at Columbia. I have seen no documentary evidence that indicates exactly when he abandoned S3, but we know that he did do so by the time he wrote *Symbolic Logic* with Harold Langford. In *Symbolic Logic*, he accepts S2 over S3. The reasons for this change are discussed in section 2.6 below. In Appendix II of *Symbolic Logic*, Lewis offers five alternative logics, S1-S5, but explicitly rejects S5 as a basis of a theory of inference. In letters, Lewis made it clear that his view of S4 was similar to his view of S5. Although he explicitly rejects S3 as well in *Symbolic Logic*, as late as the mid-1940s he considers bringing it back as his central logical system.[4] S1 is offered as a fallback system, to be adopted in case S2 turns out to have similar failings as S3. What is it about S1, S2, and S3 that appeals to Lewis?

In order to understand Lewis's positive attitude towards these logics, we need to know what these logics are.

Lewis takes possibility, conjunction, and negation as his primitive connectives. I use $\Diamond$, $\wedge$, and $\neg$ for these. He defines disjunction using the standard DeMorgan definition, material implication in a standard way, and defines strict implication as follows:

$$A \strictif B =_{df} \neg \Diamond (A \wedge \neg B)$$

Following Lewis, I use '=' as strict equivalence. Lewis does not define necessity, but I use the standard $\Box$ for his $\neg \Diamond \neg$.

Here is the array of principles that Lewis gives in (Lewis and Langford, 1951, p 493):

---

[4]In a draft of Lewis (1951).

2.5. JUSTIFYING LOGICAL PRINCIPLES

A1 $(p \land q) \strictif (q \land p)$ B1 $(p \land q) \strictif (q \land p)$
A2 $(q \land p) \strictif p$ B2 $(p \land q) \strictif p$
A3 $p \strictif (p \land p)$ B3 $p \strictif (p \land p)$
A4 $(p \land (q \land r)) \strictif ((p \land q) \land r)$ B4 $(p \land (q \land r)) \strictif ((p \land q) \land r)$
A5 $p \strictif \neg\neg p$ B5 $p \strictif \neg\neg p$
A6 $((p \strictif q) \land (q \strictif r)) \strictif (p \strictif r)$ B6 $((p \strictif q) \land (q \strictif r)) \strictif (p \strictif r)$
A7 $\neg\Diamond p \strictif \neg p$ B7 $(p \land (p \strictif q)) \strictif q$
A8 $(p \strictif q) \strictif (\neg\Diamond q \strictif \neg\Diamond p)$ B8 $\Diamond(p \land q) \strictif \Diamond p$
                                                            B9 $\exists p \exists q (\neg(p \strictif q) \land \neg(p \strictif \neg q))$

S3 is given by A1-A8, together with the rules of replacement of adjunction, strict equivalents, uniform substitution and modus ponens for strict implication. S2 is given by B1-B9 together with those rules and S1 is B1-B7 and B9 with the same rules.

B9 is a very strange axiom. It is discussed at length in (Lewis and Langford, 1951, ch 6 §6). It is not an axiom in the usual sense. It is not in the object language. It includes quantifiers which are not axiomatized in the logics. It is really a statement in the semantic metalanguage saying that the logic does not have the two-element boolean algebra as a model. Lewis wants this axiom to distinguish strict implication from material implication. Thus, although in a straightforward way S1 and S2 are subsystems of classical logic (just add that every diamond formula is equivalent to the same formula without a diamond in front), from a semantic point of view those logics are not subsystems: some classical models are barred as models for S1 and S2 (even though they are, in the normal sense, models for S1 and S2). In what follows, however, I ignore B9, since its acceptance has no affect on the list of propositional (i.e. non-quantified) theorems of the logics.

What is most striking, from a post-Kripkean point of view, about S1-S3 is that they are non-normal modal logics. In particular, they do not include the necessitation rule, nor is this rule admissible in them. Many logicians have claimed this is a flaw of these systems, causing them to have inelegant model-theoretic semantics. Moreover, Lewis's failure to discover the rule of necessitation (or, perhaps, his failure to distinguish it from the 4 axiom, $\Box p \strictif \Box\Box p$) caused him to fail to notice the logic T, and the formulation of T had to wait until Robert Feys published it in 1937 Feys (1937). From the point of view of system closure, however, the omission of necessitation makes good sense. Lewis's logics are themselves systems. They are closed under conjunction and provable modus ponens. They are theories of logic. Not all systems are about logic and so they do not contain all the theorems of logic. Lewis suggests in his 1923 lecture notes that the meaning of $\neg\Diamond\neg A$ (i.e. $\Box A$) is that A is included in all systems. In order to bar the theorems of logic from most systems, Lewis's logics reject the rule of necessitation.

2.5 Justifying Logical Principles

On the justification of logical principles, Lewis says that

For the trains of reasoning in these proofs we must, of course, appeal to intuition. A point of logic being in question, no other course is possible. But let the reader ask himself whether they involve any mode of inference which he is willing to be deprived of – for instance, in making deductions in geometry. (Lewis and Langford, 1951, p 252)

Here Lewis is appealing to the principles that we need for the closure of certain systems – geometrical theories – in order to justify logical principles.

Among those inferences that one would want when doing geometry are all the inferences licensed by classical propositional logic. S1-S3 all are closed under the rule,

$$\frac{\vdash_{PC} A}{\vdash \Box A}$$

I call this rule 'NPC'. Edward Lemmon uses NPC in his axiomatisation of S1-S3 Lemmon (1957), and Lewis was clearly aware that his logics were closed under it. NPC entails that $((A \supset B) \land A) \rightarrow_3 B$ is a theorem of S1-S3, as well. Thus, all systems are closed under all classical inference rules.

In order for NPC to ensure that all populated systems contain all the theorems of classical logic, we need to know that $\vdash_{Sn} \Box A$ entails that A is in every system of Sn (for $n \in \{1,2,3\}$). The rule that is required, is what I call 'necessary inclusion' or 'NI':

$$\frac{\vdash \Box A}{\vdash B \rightarrow_3 A}$$

NI is derivable in each of S1-S3. This was proven by Soren Halldén in 1948 Halldén (1948). Here is a slightly modified and abridged version of his proof:

Proposition 2.5.1. *It is provable in S1 that* $\Box p \supset (q \rightarrow_3 p)$.

Proof. By NPC, it is provable in S1 that

$$(1)\ \neg p = (\neg p \land (q \lor \neg q)).$$

So, by replacement for strict equivalents we get

$$(2)\ \neg\Diamond\neg p = \neg\Diamond(\neg p \land (q \lor \neg q)).$$

Moreover, by the definition of strict implication and propositional calculus, $\neg\Diamond(\neg p \land (q \lor \neg q)) = (\neg p \rightarrow_3 (q \land \neg q))$. And, by axioms B1 and B2, $(q \land \neg q) \rightarrow_3 \neg q$. Thus, we obtain

$$(3)\ \neg\Diamond\neg p \supset ((\neg p \rightarrow_3 (q \land \neg q)) \land ((q \land \neg q) \rightarrow_3 \neg q)).$$

By a bit of PC, (3), axiom B6 and modus ponens, it is derivable that

$$(4)\ \neg\Diamond\neg p \supset (\neg p \rightarrow_3 \neg q).$$

And $(\neg p \rightarrow_3 \neg q) = (q \rightarrow_3 p)$, so (4) entails

$$(5)\ \neg\Diamond\neg p \supset (q \rightarrow_3 p).$$

That is

$$(6)\ \Box p \supset (q \prec p).$$

□

Halldén proves a slightly stronger theorem than NI, but it clearly entails NI, since the theorems for each of S1-S3 are closed under modus ponens for material implication.

Halldén treats the provability of $\Box p \supset (q \prec p)$ as a negative property of S1. It is a paradox of strict implication. Halldén has a point – this thesis is very closely related to those propositions that McColl and Lewis used to criticise material implication. I am not sure, however, that by the 1920s Lewis would have been bothered by it. In the notes for his 1923 lectures on logic at Columbia University, Lewis writes the following:

Consistency and "States of Affairs" or "Systems".

A prop[osition] is possible ~~or self-co[nsistent]~~ if it is true in some system or state of affairs.

A prop[osition] is necessary if it is true in every system.

Two prop[osition]s are jointly consistent if they are jointly true in some system.

$P \prec Q$ if Q is in every system that P is in.

If we take the 'if's in this passage to be mathematician's 'if's (i.e. as having the force of 'iff's), then these statements seem to set out truth conditions for the modal connectives. Putting together the conditions for necessity and strict implication, one can immediately derive that $\Box p \supset (q \prec p)$.

One might object that Lewis could have refrained from giving truth conditions for the modal connectives in terms of systems and yet retained the view that systems are logically closed. In addition, he could have, perhaps, avoided the paradoxes of strict implication but still maintained that if $\Box A$ is a theorem that A is in every system. But having $\Box p \supset (q \prec p)$ in the logic forces NI to be valid and so allows Lewis to state a more succinct and more clearly motivated definition of a system. This gives Lewis's theory of systems more theoretical simplicity and greater unification. For a pragmatist, that is extremely important.

2.6 Transitivity of Strict Implication

As we saw in the previous section, Lewis claims that we should accept whatever principles we cannot imagine doing without when doing deductions in actual systems. In doing deductions in geometry and other mathematical theories, lemmas are proven from axioms and definitions, and then theorems are derived from the lemmas. Hence the theorems are derived from axioms. This procedure requires that the relation of deducibility be transitive. In terms of entailment, it would seem that one of the following two theses be valid:

$$(WT)\ ((A \prec B) \wedge (B \prec C)) \prec (A \prec C)$$

$$(ST)\ (A \prec B) \prec ((A \prec B) \prec (A \prec C))$$

'WT' stands for 'weak transitivity' and 'ST' stands for 'strong transitivity'. WT is just axiom A6 (=B6) and is sometimes called 'non-nested transitivity' and sometimes 'conjunctive syllogism'. ST is sometimes called 'nested transitivity' and sometimes 'suffixing'.

By 1932, Lewis rejects ST in favour of WT. Here is perhaps Lewis's only statement of what he thinks is wrong with ST:

> I doubt whether this principle should be regarded as a valid principle of deduction: it would never lead to any inference $p \prec r$ which would be questionable when $p \prec q$ and $q \prec r$ are given premises; but it gives the inference $(q \prec r) \prec (p \prec r)$ whenever $p \prec q$ is a premise. Except as an elliptical statement for "$((p \prec q) \wedge (q \prec r)) \prec (p \prec r)$ and $p \prec q$ is true," this inference is dubious. (Lewis and Langford, 1951, p 496)

I think that it helps to understand what Lewis is saying by making his point in terms of systems. Suppose that in a physical theory, t, $p \prec q$ is a law. If the theory is closed under ST, then $(q \prec r) \prec (p \prec r)$ is also law of t. Now, suppose that m is a subordinate system of t, say a model of t. Then $(q \prec r) \supset (p \prec r)$ is in m. Let's suppose that $p \prec q$ is not in m. Then we can understand m's containing $(q \prec r) \supset (p \prec r)$ by relating m to t and recalling that $p \prec q$ is in t. $(q \prec r) \supset (p \prec r)$ is an isolated and, seemingly irrelevant, state of affairs in m.

There might seem, however, to be a similar problem with S2. The rule form of ST (RST) is derivable in S2:

$$\frac{\vdash A \prec B}{\vdash (B \prec C) \prec (A \prec C)}$$

Whether Lewis knew that RST is derivable in S2, and if he did, when he knew it, is unclear. It can be derived from what Hughes and Cresswell call (Hughes and Cresswell, 1968, p 231) call *Becker's Rule* (BR):

$$\frac{\vdash A \prec B}{\vdash \Box A \prec \Box B}$$

In a paper published in 1951, Lewis says that BR is derivable in S2 (Lewis, 1951, p 428) and he presents a proof for it in Appendix III to the second edition of *Symbolic Logic* (Lewis and Langford, 1959, pp 507-508) (see also (Hughes and Cresswell, 1968, p 232)). I doubt that Lewis knew of the derivability of Becker's rule in S2 before the late 1940s. In a draft of Lewis (1951), he considers adopting S3 as the basis for his logic of properties because it contains axiom A8 (the thesis form of BR). But he abandons

2.6. TRANSITIVITY OF STRICT IMPLICATION

S3 in the final version of Lewis (1951) because he realises then that the rule form is derivable in S2.

Here is a sketch of a proof of RST in S2.[5]

Proposition 2.6.1. *The following is a derived rule of S2:*

$$\frac{\vdash A \strictif B}{\vdash (B \strictif C) \strictif (A \strictif C)}$$

Proof. Assume the premise:

$$(1) \vdash A \strictif B$$

By NPC,

$$(2) \vdash (A \supset B) \strictif ((B \supset C) \supset (A \supset C)).$$

From (2) and Becker's rule, we can then derive

$$(3) \vdash (A \strictif B) \strictif ((B \supset C) \strictif (A \supset C)).$$

From (1) and (3) and modus ponens we get

$$(4) \vdash (B \supset C) \strictif (A \supset C)$$

By (4) and Becker's rule again, we derive

$$(5) \vdash \Box(B \supset C) \strictif \Box(A \supset C),$$

that is,

$$(6) \vdash (B \strictif C) \strictif (A \strictif C),$$

as required. □

Fortunately, RST is not as problematic for Lewis as ST is. Suppose that we define '$bachelor(x)$' as '$man(x) \land \neg married(x)$'. Thus, $bachelor(i) \strictif \neg married(i)$, for all i, is a theorem of S2 with this definitional extension. Hence

$$married(i) \strictif \neg bachelor(i)$$

is also a theorem. Suppose that we want to formalise the rules for a reality television show in which the central character is a bachelor, so we have:

$$central\ character(i) \strictif bachelor(i)$$

In the system of rules for the show. Thus, since tis system is closed under the theorems of the definitional extension of S2, we also have

$$(\neg bachelor(i) \strictif \neg central\ character(i)) \supset (married(i) \strictif \neg central\ character(i))$$

in the system, and this seems innocuous (and correct).

---
[5]This proof, in more or less the current form, is due to Max Cresswell.

2.7 The Consistency Postulate and S1

Lewis says that he is willing to abandon the consistency postulate and adopt S1 if it turns out that ST is a theorem of S2. To remind the reader, the consistency postulate is

$$\Diamond(p \wedge q) \strictif \Diamond p.$$

Lewis also defines a consistency operator $\circ$:

$$p \circ q =_{df} \neg(p \strictif \neg q)$$

(Lewis and Langford, 1951, 17.01). Thus, p and q are consistent with one another if and only if p does not entail the negation of q. By the definition of strict implication, then,

$$(p \circ q) = \neg\neg\Diamond(p \wedge \neg\neg q).$$

By PC, and the replacement of equivalents:

$$(p \circ q) = \Diamond(p \wedge q)$$

Thus, the consistency postulate can be also read as

$$(p \circ q) \strictif \Diamond p.$$

This can be read as saying that if p is consistent with some proposition, then p is consistent with itself.

To abandon the consistency postulate to avoid ST seems, at first glance, to be extremely rash, bordering on silly. The truth condition for consistency given in section 2.5 says that two propositions are consistent with one another if they are both in the same system and the truth condition for possibility says that a proposition is possible if it is in some system. So, the consistency of two propositions seems to entail the possibility of both. When the consistency postulate is abandoned, the operators $\Diamond$ and $\circ$ seem to have merely technical uses and no real meaning.

I suggest that, for Lewis, at least in the 1920s and early 1930s, the need for a coherent theory of meaning for the modal connectives is less important than having a good theory of system closure. Treating the possibility and consistency connectives as mere technical devices is in keeping with this.

Lewis's work from the late 1930s until the end of his life, beginning with his and Langford's 1935 manuscript "A Note on Strict Implication" Lewis and Langford (2014), shows a much greater concern with finding a theory of meaning for the modal connectives. The start of his work on the theory of meaning for the modals coincides with William Parry's proof that A8 and ST are not theorems of S2 Parry (1934). After this, it seems, Lewis is fairly firm on his choice of a logic of strict implication and is able to accept the consistency postulate. This enables him to begin to develop the semantics of intensions that we see in his work of the 1940s Lewis (1946, 1951).

2.8 What Lewis Entailment is Not

Lewis's claim that strict implication represents deducibility has been widely misunderstood. Ruth Barcan Marcus says:

> it is plausible to maintain that if strict implication is intended to systematize the familiar concept of deducibility or entailment, then some form of the deduction theorem should hold for it. (Marcus, 1953, p 234)

The idea here is that if the Lewis systems represent deducibility, then they should represent *their own* deducibility relations. Thus, one would expect that if B is deducible from the premises $A_1, ..., A_n$ and A, that we could deduce from $A_1, ..., A_n$ that $A \prec B$. The view that a modal logic should represent its own deducibility relation has its most famous expression in Dana Scott's "Engendering an Illusion of Understanding" Scott (1971). In that paper, Scott argues that S4 can be understood in that manner.

But, as Barcan Marcus shows, S1-S3 fail to satisfy the deduction theorem. For S2, her example is one that we have already examined. Becker's Rule is derivable in S2, and so $\Box A \prec \Box B$ is derivable from $A \prec B$, but it is not the case that axiom A8 of S3 is provable in S2. Thus, the deduction theorem fails in S2. Barcan Marcus concludes that the strict implication of S2 does not represent deducibility.

I have suggested that for Lewis, B is deducible from A if and only if all systems that contain A also contain B. Logics are systems too, but just as most other systems are not merely closed under logical laws, logics are closed under laws that are themselves not represented in the logic itself. For example, the Lewis systems are all closed under uniform substitution for propositional variables. This is not represented in the logic (or any useful logic) by a strict implication.[6] Most systems that are not logics, however, are not closed under uniform substitution. The specific content assigned to propositional variables makes a difference in most of the theories that we hold. Thus, one would not expect that a logic that represents entailment in Lewis's sense would represent the specific rules of deducibility in the logic itself.

2.9 Concluding Remarks

I have suggested that Lewis understands his logics, during the 1920s and early 1930s at least, primarily as theories of theory closure. The way in which he describes the intuitions one should use to justify principles of logic support this suggestion. He says that we should think of those principles that we would not want to do without when reasoning about theories, such as geometries, when deciding what postulates and rules should be included in a logic. In a way Lewis's attempt to construct a theory of theory closure is not very ambitious. A theory that represents its own deducibility relation is much more difficult to construct. But it is also a very different project. Confusing the two has led, in my opinion, to some unwarranted criticisms of Lewis.

---

[6]In a sense it is represented in S4 and S5 in the sense that if A and B are theorems and B results from the uniform substitution for some propositional variables in A, then $A \prec B$ is also a theorem. But the way in which Scott and others formulate the representation does not assume that antecedents are theorems, so in that sense uniform substitution is not represented in those logics.

Reading Lewis's logical writings together with his epistemological and metaphysical writings yields a much different picture of Lewis's logic than reading the logical works alone. I recommend to anyone studying Lewis to look at his work in this wider context.

Acknowledgements: I am grateful to the archivists at Stanford University Library and at Harvard University for helping me to find material in the Lewis, Quine, and Royce archives. I read an early version of this paper at the 2011 Asian Logic Conference/Australasian Association of Logic Conference in Wellington. I am especially grateful to Max Cresswell, Nick Smith, and of course Peter Schotch for discussions related to this paper. The research for this paper was funded by the Marsden Fund of the Royal Society of New Zealand.

2.10 Bibliography

Feys, R. (1937). Les logiques nouvelles des modalités. *Revue Néoscolastique de Philosophie 40*, 517-553.

Goheen, J. and J. Mothershead (Eds.) (1970). *Collected Papers of Clarence Irving Lewis*. Stanford, CA: Stanford University Press.

Halldén, S. (1948). A note concerning the paradoxes of strict implication and Lewis's system S1. *The Journal of Symbolic Logic 13*, 138–139.

Hughes, G. and M. Cresswell (1968). *An Introduction to Modal Logic*. London: Methuen.

Lemmon, E. (1957). New foundations for Lewis modal systems. *The Journal of Symbolic Logic 22*, 176–186.

Lewis, C. (1918). *Survey of Symbolic Logic* (first ed.). Berkeley: University of California Press.

Lewis, C. (1920). Strict implication – an emendation. *The Journal of Philosophy, Psychology and Scientific Methods 17*, 300–302.

Lewis, C. (1921). The structure of logic and its relation to other systems. *The Journal of Philosophy 18*, 505–516. Read at the conference of the American Philosophical Association, December 1920. Reprinted in (Goheen and Mothershead, 1970, pp 371-382). Page references are to the reprinted version.

Lewis, C. (1923a). Facts, systems, and the unity of the world. *The Journal of Philosophy 20*, 141–151. Reprinted in (Goheen and Mothershead, 1970, pp 383-393). Page references are to the reprinted version.

Lewis, C. (1923b). A pragmatic conception of the *a priori*. *The Journal of Philosophy 20*, 169–177. Reprinted in (Goheen and Mothershead, 1970, pp 231-239). Page references are to the reprinted version.

2.10. BIBLIOGRAPHY

Lewis, C. (1929). *Mind and the World Order: Outline of a Theory of Knowledge*. New York: Charles Scribner and Sons.

Lewis, C. (1932). Alternative systems of logic. *The Monist 17*, 481–507. Reprinted in (Goheen and Mothershead, 1970, pp 400-419). Page references are to the reprinted version.

Lewis, C. (1946). *Analysis of Knowledge and Valuation*. LaSalle, IL: Open Court.

Lewis, C. (1951). Notes on the logic of intension. In H. Kallen and S. Langer (Eds.), *Structure, Method and Meaning: Essays in Honor of Henry M. Sheffer*, pp. 25–34. New York: Bobbs-Merrill. Reprinted in (Goheen and Mothershead, 1970, pp 420-429). Page references are to the reprinted version.

Lewis, C. and C. Langford (1951). *Symbolic Logic* (first ed.). New York: Dover. Originally published in 1932.

Lewis, C. and C. Langford (1959). *Symbolic Logic* (second ed.). New York: Dover.

Lewis, C. and C. Langford (2014). A note on strict implication. *History and Philosophy of Logic 35*, 44–49. The manuscript of this article was written before or in 1935.

Marcus, R. B. (1953). Strict implication, deducibility, and the deduction theorem. *The Journal of Symbolic Logic 18*, 234–236.

Parry, W. (1934). The postulates for 'strict implication'. *Mind 48*, 78–80.

Scott, D. (1971). On engendering an illusion of understanding. *Journal of Philosophy 68*, 787–807.

Chapter 3

The Logical Project

Letitia Meynell[1] and Richmond Campbell[2]

3.1 Introduction

Those of us trained in the analytic tradition who are not logicians often find ourselves oddly positioned with respect to logic. On the one hand, we are convinced of the use and power of formal logic. We are comfortable translating the occasional sentence or argument into predicate or modal form and impressed by the clarity it can provide. We gladly tell our students that large swaths of 20th century philosophy will make no sense to them unless they have mastered basic logic. On the other hand, we also find that in much of our own philosophical practice our arguments are more inductive in character and typically fail to fit any neat argument form in any current logical system. Furthermore, once we journey beyond predicate logic, modal logic, and basic set theory into the mysterious realms of metalogic we are ill at ease. The rigor, clarity and inexorability that we so value in formal logic, metamorphoses into something that seems more like a sacred language, comprehensible only to the initiated few who are blessed with the gift of insight. We understand that metalogic is the foundation upon which the formal edifice sits, but the controversial character of metalogic and the ease with which new rules are proposed or dismissed seems to undermine the rigor and clarity that we so prize in formal logic in our own philosophical practice. We start wondering what exactly this logic stuff is anyway—a question we dare not ask lest we out ourselves as impostors and are dismissed from the analytic fold as heretics. (This feeling of courting heresy and banishment is particularly acute for those of us who are naturalized, let alone feminist, as we are already viewed with some suspicion by many in the analytic mainstream.)

It is in this uneasy predicament that we find Peter as a champion. First of all, Peter stands out as a logician who firmly believes that we ought to be able to do things with formal systems, as seen in his work with David Braybrooke (Braybrooke, Brown and

---

[1]Department of Philosophy, Dalhousie University, letitia.meyell@dal.ca
[2]Department of Philosophy, Dalhousie University, richmond.campbell@dal.ca

Schotch 1995). Secondly, he asks our heretical question—what is logic?—and thinks that we should expect logicians to have an answer (Schotch Forthcoming, Ch. 2). Thus a celebration of Peter's work seems like one of the few safe places to explore this question as interested outsiders. We are empowered by Peter to ask the question and our answer is informed by Peter's work, even as it predominantly reflects preoccupations of our own. We will suggest that logic is just one of a set of evolving human 'projects,' all of which are rooted in the social character and cognitive capacities of our species. The Logical Project names our collective effort to figure out how to reason well. In this way we offer a kind of naturalized account of logic that places its origins and function in an evolutionary context. One of the key concerns with this approach is how a descriptive empirical account can ground the normativity of logic. In other words, how can our application of logical systems and methods be considered justified without some kind of external logical facts justifying them? In informal logic this is, in effect, the problem of induction, though, as we will see, there is a similar problem for formal logic when we think of it as a changeable product of cultural evolution (Goodman 1955).

We propose to proceed as follows. In the next section we identify the function of logic in the normative sense of identifying what logic is for. Here we follow Peter's refreshingly straightforward account, identifying logic with the theory of inference (Schotch Forthcoming, 4 ff.).[3] The study of logic identifies the character and form of good inferences. We then turn to the grandfather of contemporary naturalized philosophy, W.V.O. Quine, and consider his remarks on the evolutionary history of induction. We suggest that the practice of informal logic—in particular, induction in a narrow sense—reaches back in time, deep into our evolutionary past, and is vindicated by its survival value, as Quine suggested. The evolved biological function of our inductive capacity is to make good inferences. We then address the key objections that this idea incites. These are familiar philosophical concerns about the justification of induction—circularity, normativity and incompleteness. Even as we meet these objections, basic evolutionary theory can only take us so far, giving us no more than a rough and ready explanation and justification of very basic inductive capacities that are presumably shared across a wide variety of cognitive animals. Moreover, it provides only limited insight into how these evolved capacities are related to contemporary inductive projects, such as the natural sciences, let alone deduction. To fill out the connection between these evolutionarily ancient inductive capacities and the current plethora of sophisticated and complex logical systems and methods (both inductive and deductive) we turn to Phillip Kitcher and Graham Priest.

In *The Ethical Project* (2011), Kitcher makes a case for a specific kind of cultural-historical process that produced the sophisticated moral norms that govern behaviour today out of rudimentary evolved pro-social traits. This ethical project is not only structurally analogous to the Logical Project but would have been historically preceded by it, then run parallel to it *ex hypothesi*. We envision here the two projects informing

---

[3] Not everyone shares this perspective. W.V.O. Quine believes that logic is about valid logical form; however, logic thus conceived sheds enormous light on valid inference. Michael Resnik says that logic is not about good inference so much as resolving theoretical puzzles, such as about quantum physics, Hilbert space, or metaphysical modes of being (Resnik 1985, 234). We would counter that logic is still about good inference generally, including how to reason well about these rarified subjects.

3.2. WHAT IS LOGIC?

each other just as the two evolutionary processes—cultural and biological—run parallel and also interact. We take from Kitcher's cultural-historical account some of the basic mechanisms of the cultural evolutionary process that explain how we developed contemporary logic. Ultimately, we hope to show that not only formal logic but also sophisticated inductive reasoning of the kind that is associated with the natural sciences developed from, but are irreducible to, our evolutionarily ancient, untutored inductive capacities. Language plays a crucial role here. For contemporary rarified logical projects, such as natural deduction or statistical analyses of data, only beings with sophisticated linguistic capacities are capable of employing them or deliberating about them, even if other forms of logical inference are available to non-linguistic beings (human and non-human).[4]

We find that the problem of normativity once again arises when we take this cultural evolutionary approach to logic. To address it we employ a useful recent discussion by Graham Priest about the rational revision of logic (2014). Priest distinguishes various different things we might mean when we ask whether logic can be rationally revised. His analysis is extremely sensitive to the history of logic—which is, in effect, a history of revisions—and the real world uses of logic. His focus on the historical facts, actual practices, and real world applications makes Priest's approach a natural fit for a naturalized theory of logic. Ultimately, we argue that induction and deduction are two interrelated strands of the Logical Project—*our collective effort, stretched out through our history and natural history, to figure out how to make good inferences.*

This account has all the failings of any other evolutionary Just-So story, being highly speculative and necessarily under-evidenced. Nonetheless, it seems to us that something like this is likely true and we hope that others, who are taken with the general idea and more knowledgeable of comparative psychology and the global history of logic, may fill in the details with better-evidenced, less anthropocentric, and less Eurocentric accounts. In this essay we do not intend to do science, that is, to attempt to marshal all the empirical evidence needed to back up our narrative. Instead our object is to sketch a plausible narrative that makes sense of logic as an evolving human (and possibly nonhuman) project designed to meet our needs and interests and that may inspire empirical research that could fill in and give substance to the picture we draw.

3.2 What is Logic?

3.2.1 A Theory of Inference

Logic is the study of inference (Schotch Forthcoming, 4 ff.). Even if contemporary logicians have attempted to pursue their discipline as a formal system detached from any application, the historically naked function of logic has been to articulate, examine, and defend rules of inference. We value logic and apply its methods to our most important epistemological enterprises because through using it we make better inferences. This

---

[4] It is in treating the evolution of induction and deduction as in an important sense one multifaceted project and in making cultural as well as biological evolution central that our account differs most from Schechter (2013).

appears to be true of both informal and formal logic, even as the character of the inferences made in each realm differ in terms of their certainty and ampliative capacity. The marvel of formal logic is truth preservation—the remarkable fact that with certain forms of inference if one starts with true premises one necessarily ends up with a true conclusion, regardless of the actual content of the sentences. Indeed, as Peter reminds us, truth preservation defines formal validity (Schotch Forthcoming, 12). Informal logic still interrogates inference, but only those patterns of reasoning that are truth-conducive, not truth-preserving. (For the purposes of this paper we will treat informal logic as roughly synonymous with inductive logic, despite the unfortunate implications for inductive formulae, e.g., probabilistic reasoning.)

It is worth remembering that inference and argumentation are logically (as opposed to psychologically) one and the same. As Brendan Gillon succinctly puts it: 'Humans reason: that is, taking some things to be true, they conclude therefrom that other things are also true. If this is done in thought, one performs an inference; and if this is done in speech, one makes an argument. Indeed, inference and argument are but two sides of the same coin: an argument can be thought, and hence become an inference; an inference can be expressed, and hence become an argument' (2011, §1). Inference, moreover, is a *cognitive* kind so that even if it does not require language it does require something like internal representational states. Thus it is not enough to identify behaviors that can be given an explanation at an ultimate level following a deductive form. Inferences are cognitive processes that at the proximate level reveal an inductive or deductive form.

If logic is the study of inference, then the questions for the naturalized logician are, what is the evolutionary history of those inferential forms, what are the capacities that ground them, and can this evolutionary history explain their normative force? Needless to say, these challenging questions have a significant empirical component, which in the end is best left to comparative psychology and ethology (cognitive and otherwise). Nonetheless, philosophers have had something to say on the matter, starting with W.V.O. Quine. Quine shows us how this normative function of logic may also be its function in the biological sense by pointing to the selective advantage of making good inferences.

3.2.2 An Evolutionary Theory of Inductive Inference

Induction at its most basic level is found in the tendency, presumably across multiple taxa, to infer future similarities from past similarities. Is this tendency worth having? That depends on whether the inferences based on this tendency are any good, given that we care about our survival and moderately good inferences are a necessary means to survival. Quine puts it this way, 'For me ... the problem of induction is a problem about the world: a problem about how we, as we now are (by our present scientific lights), in a world we never made, should stand better than random or coin-tossing chances of coming out right when we predict by inductions which are based on our innate, scientifically unjustified, similarity standard' (Quine, 1969, 127). Given the connection between good inferences and survival, the implication would appear to be that our evolutionarily ancient, untutored inductive tendency[5] is worth having and

---

[5] What exactly do we mean by 'our evolutionarily ancient, untutored inductive tendency'? We shall leave this question largely unanswered since our thesis does not depend on making the reference precise, but we

3.2. WHAT IS LOGIC?

indeed justified if, but only if, the tendency enables us to do better than chance in making true (or true enough) predictions.

Quine famously applies evolutionary theory to argue that the condition is met. Natural selection, he notes, provides a partial explanation of why our untutored inductions actually do better than chance at predicting the future: if they did not, the ancestors from which we have inherited our similarity standard[6] would not have survived to reproduce their kind (Quine, 1969, 126).[7] It follows that evolutionary theory, if true, would *justify* our tendency to make inductions based on this inherited similarity standard. Why? Because, in so far as the evolutionary explanation is successful in explaining why such inductions do better than chance at predicting the future, it gives us *ipso facto* sufficient reason to believe that these inductions do fulfill their function, hence are epistemically worthwhile, and otherwise would not exist. The Darwinian explanation of the practice, if true, thus provides an *instrumental* justification of it by showing how and why it succeeds in fulfilling its goal or function. More than that, the fact that our inductive tendency does better than chance in leading to true predictions is the evolutionary reason why this tendency exists. The reason why it exists and the reason why its existence is justified are, in a word, *identical* (Cf. Campbell, 1998, 87-95).

It may be objected that to describe our tendency to make inferences about the future, along with the causal origins of that tendency, is not in itself to make a normative claim about whether the tendency is worth having or, for that matter, whether the inferences that result are justified.[8] Such normative claims would purport to tell us how we *ought* to think; a claim based on Darwinian ideas about the causal origins of our inductive tendency, however, tells us at best only how we *do* think. The difference between the two means that the Darwinian story cannot be a *justification* of our tendency to infer future similarities from past ones. But this objection from normativity fails because the instrumental justification described in the previous paragraph contains a normative

---

take it to be roughly the collective 'native inferential tendencies' that began to be studied in the early 1970s by researchers like Daniel Kahneman and Amos Tversky, albeit with corrections accounting for developmental plasticity. Our view of this 'native' inductive ground is more positive than theirs and is captured best by Hilary Kornblith (1993, 83-107). Like him, we take the untutored tendency to work well when applied in situations like those in which it evolved (which is not to rule out the possibility of it also working well in completely novel environments, i.e., exaptations). We are indebted to Austin Booth for pressing us on this point.

[6] We do not mean to commit ourselves to a theory of innate ideas here. For the sake of this argument it is enough that this similarity standard rests on cognitive mechanisms (some of which may be content neutral, some of which may be content specific) that in the environments inhabited by our ancestors tended to produce a similar enough similarity standard to be 'visible' to natural selection.

[7] It is worth noting that although this is a natural selection account it is far from adaptationist. Doing better than chance is a very low bar and allows for a variety of cognitive mechanisms (some of which are, presumably, narrowly focused on a specific function; others of which are, presumably, general purpose). We do not have the view that there is a finely-tuned optimal inferential module or anything of the sort. We are only committed to the idea that a basic inductive tendency if it were somewhat successful at getting the world right would be selected for. The precise mechanisms underlying this capacity are presumably variable among species in terms of their character and degree of success. While some may have had a long history of information processing, even prior to development of a properly inferential capacity, others may well be spandrels.

[8] The two normative claims are connected: the inferences that result from the tendency are justified when they do better than chance at getting it right (an end we value) and the tendency is worth having for a further reason if in giving us inferences that are justified, it helps us to survive (another end we value).

premise, namely that the goal or function of the natural tendency—for us to do better than chance at making true predictions—is worthwhile. The premise can be challenged, of course. One might question whether there is value in correctly predicting future outcomes, though it is safe to suppose that most people think there is value. The point here, however, is simply that, whether we question it or not, the premise is fully normative. One might doubt that Quine would have conceded that his explanation of the success of our inductive tendency could properly be called a *justification* of its existence. But our aim is not to give an exegesis of Quine's thought. The problem of explaining how we do better than chance by relying on our natural inductive tendency is interesting on its own since doing better than chance matters to most of us. Because it does matter to us, the value that we see in getting it right about the future provides the unstated normative context in which the problem of induction is posed and indeed the reason why the problem of justifying induction demands our attention. We want to know how and why we can attain this valuable end when we rely on our natural inductive tendency. If the explanation Quine gives is right, we have an answer that instrumentally justifies the tendency, provided only that the aim that Quine identifies is worthwhile.

3.2.3 Some Clarifications

A few caveats are in order, addressing both the content of these inductions and the concept of justification itself. First, it is possible to use the word 'justify' thinly to signify being an efficient means to an end. Thus one might judge that someone would be justified in ingesting lots of arsenic if her only aim is to kill herself, without passing any judgment on whether her goal is worthwhile. If in the context of 'justifying' the use of our inductive tendency one intends only this thin sense of justification, then there is clearly no need to suppose an unstated normative assumption that the end is worthwhile. But, notice, by the same token the objection from normativity would have no target. Our thesis is simply that when 'justify' is used in the thick sense, as it normally is when philosophers talk about justifying induction, there is no problem of normativity either, since there is an unstated normative premise.

The next caveat is that 'function' can be ambiguous when talking about natural selection explanations for why something exists. In many biological contexts, what is meant by 'function' is roughly what has come to be called the 'etiological' or 'selected effect' use of the term: The function of X is Y only if the existence and/or maintenance of X is to be explained through selection of Y (Kitcher, 1993, 265).[9] We believe that this use of the term does not imply the normative premise in question, and we grant that the selection story for the case at hand implies that the inductive tendency

---

[9] For those readers less familiar with evolutionary reasoning it is perhaps useful to fill in X and Y for the Quinean induction case: The function of the evolutionarily ancient, untutored inductive tendency is to produce better than chance inferences only if the existence and/or maintenance of the evolutionarily ancient, untutored inductive tendency is explained by selection for the production of better than chance inferences. Readers are reminded that selection is stochastic, so that the innumerable counterexamples where 'good' inductions will have led to evolutionary bad consequences for our ancestors and kin are insufficient to ground rejection unless it can be shown that these types of event have been in sum more deleterious to reproductive success than those cases where 'good' inductions helped their agents thrive.

3.2. WHAT IS LOGIC? 47

has the function noted earlier in roughly this biological sense. We maintain, though, that over and above the biological claim about the function of the tendency there is the normative assumption that this biological function is a *worthwhile* effect and that the latter assumption blocks the normative objection. This is in effect to say that the biological function of our inductive capacity is also normatively functional in the sense laid out in section 3.2.1 above. We have evolved to make good inferences; indeed, we study logic because we value making good inferences and hope it will help us make better ones.

Third, even if Quine does not advert to the fact, the inductive tendency just described seems to go deep into our evolutionary history and is likely shared not only with other primates but with a good many other animals too. It is reasonable to suppose that there are multiple different cognitive mechanisms that can instantiate this inductive tendency and many other cognitive and affective capacities besides that accompany them.[10] For all cognitive mechanisms that support inferences, Quine offers prima facie reason to suppose that natural selection will tend to favour those inferential mechanisms that tend to get things right. This allows that some inferential mechanisms might tend to produce results that rarely if ever get the world right but nonetheless serve natural selection's reproductive goals. However, it seems highly unlikely that there would ever be any type of creature endowed only with inferential mechanisms producing systematically absolutely false beliefs.[11] As Quine famously notes, 'Creatures inveterately wrong in their inductions have a pathetic but praiseworthy tendency to die before reproducing their kind' (1969, 126).

Finally, it's worth noting Quine's tendency to avoid the word 'truth,' instead characterizing the successful inferences of our ancestors as 'coming out right.' When we consider the ancient character of this basic inductive tendency this makes good sense. The use of 'true' and even 'true enough' (Elgin 2004) to refer to the 'good' inferences of non-human animals, and many humans besides, is somewhat misleading. Presumably, many of the types of inferences that are included under Quine's description are not going to have sentential or propositional content and so will not be the kinds of things we can properly think of as being strictly speaking truth functional. Instead, we assume

---

[10] This opens the tantalizing possibility that other species may have their own logical projects that are importantly structurally similar to the human one. In *Thinking Without Words* (2007), José Bermúdez explores nonhuman animal reasoning forms based on protonegation and protoconditional thinking that resemble deduction in their syntactic form. As we shall argue in section 3.3, below, this kind of logical capacity, though necessary, is not sufficient for the Logical Project to get off the ground. The Logical Project begins in reflection and deliberation on inferences. Joseph Call (2006) makes a compelling case for the metacognitive capacities of some nonhuman animals. While without language it is difficult to see how groups could engage in metacognitive deliberation, evidence of states of uncertainty and the acquisition of information to quell uncertainty suggests that non-linguistic individuals may be able to develop and improve on rudimentary norms about their own inferences, thus taking the initial steps in their own logical projects. Ultimately, it may be unreasonably anthropocentric to think that among other social, cognitively complex species there cannot be distinct, culturally inherited and evolving inferential practices which constitute other logical projects running parallel to our own.

[11] An example of an inferential mechanism that could be adaptive despite producing systematically false results might be hyper-vigilant predator detection producing numerous false positives in a prey species. If, however, the same species were also to have mate-recognition, food-recognition, poison-recognition mechanisms that produced a similar level of erroneous results one cannot imagine the species thriving.

that these basic, untutored inferences can be made with various mental representations and success can be measured by the extent to which the produced mental content is isomorphic to a set of features of the epistemic subject's environment with respect to her or his current interests.[12] This characterization allows that mental representations with non-propositional form—for instance, mental images or Tamar Gendler's aliefs (2008)—could support inductive inferences of the kind described by Quine. To regard rough isomorphisms as 'true' or even 'true enough' is to do considerable violence to the concept of 'truth.' Though these isomorphisms between non-propositional mental content and world may be consistent with the use of 'truth' among at least some epistemologists, they are poorly suited to the work of most logicians because their inherent vagueness is corrosive to truth-preservation and defies formalization. The many rough edges of these evolved inferential patterns become amenable to greater precision and truth-preservation with the advent of language.

3.2.4 Circularity and Incompleteness

It is impossible to discuss the function or inferential power of induction without addressing the famous charge of circular reasoning. Do we not assume exactly what we want to prove, namely that our basic inductions are justified, when we appeal to evolutionary theory, which is itself based in induction and ultimately based on our evolutionarily ancient, untutored inductive tendency if it is, as we argue below, a foundation for all induction? But to reason in a circle in this way undermines the validity of the supposed justification. One could say even that the problem of circularity already contains within it a problem with normativity, since by illegitimately assuming that induction is justified when Quine appeals to Darwinian theory he is at that point *illegitimately making a normative assumption*; that is, he is making a normative claim without any prior justification when such justification is needed to avoid vicious circularity.

The circularity objection can be seen to fail, however, once we distinguish between using and justifying induction. It is not unusual to find oneself engaged in an activity and then to wonder whether it is justified in this respect or that. A practice failing to be justified does not preclude it from existing, and the same is true when we engage in the unreflective practice of following our untutored inductive tendency or for that matter in reasoning inductively about whether a scientific theory is justified. To stop and ask whether one's methods of inductive reasoning are themselves justified is a further activity. That said, we will assume that it would be illegitimate if at any point in an attempt to justify induction one were to not only use induction but in addition assume that what is to be justified is in fact justified. It is critical to notice, however, that no such assumption is made or needs to be made in the appeal to evolutionary theory. In

---

[12] This roughly follows Helen Longino's generic formulation of knowledge (1994). Longino treats the content of knowledge as a model—'model' being a concept that is vague at best, as it has been defined in any number of ways by various proponents of the semantic view of theories. Models are not true or false; they are realized to some level of precision by a real world system and are in this sense isomorphic with the world. This formulation is broad enough to include a variety of more specific accounts of epistemic success, such as success semantics, which Bermúdez uses to explore the character of logical inferences among non-human animals (see Bermúdez 2007, 88 ff.).

3.2. WHAT IS LOGIC?

fact, Quine makes the last point clear when he notes that through using induction one can discover a grave problem in the appeal to evolutionary theory, such as discovering through inductive inquiry that the unreflective inductive tendency that we have been talking about fails to be correlated with genotype or some other means of heredity. In that case the attempt at inductive justification of this tendency would fail to justify it since the appeal to Darwinian theory to argue that we would not exist if the inductions did worse than chance depends for its success on the assumption of some heritable basis for the inductions.

In general, there are three levels at which the use of induction can defeat Quine's attempt to justify this inductive tendency. First, it can fail just at the level of applying selection theory to the case at hand in the way just noted or in not giving a sufficiently precise specification of what the tendency is so that it can be tested for heritability. At a second, much higher level, problems with selection theory could be uncovered through scientific (inductive) reasoning that would render Darwin's theory of natural selection defective in some general way. There has been a long history of challenges at this level. None have been successful so far but some could be at some point. At a still higher level it could turn out by using inductive reasoning that *no* scientific theory of how humans think can explain why the use of induction is or should be successful in making our predictions do better than chance at coming out right. In that dire event, as Quine concedes (Quine 1981, 22), science would fail to be able to explain itself. In particular, it would fail to justify itself using its inductive methods. Since these are all theoretical possibilities, consistent with using induction in general and our natural inductive tendency in particular, there is no vicious circle, that is, no instance in our use of induction where its use entails that it is justified.

Still, an objector might ask, does the foregoing justification of induction by appeal to Darwin differ from a straight appeal to the fact that induction has mostly worked in the past? Surely if I infer that induction is likely to work in the future from the fact that it has mostly worked in the past, I am reasoning in a circle, just as Hume famously pointed out. So what exactly is the difference? Or to put the question another way, how is an appeal to Darwinian evolution any less *viciously* circular than an appeal to the fact that induction mostly has worked in the past in order to conclude that it will likely work in the future? Note that in neither case is there any explicit assumption that induction is justified.

In our view, using the fact that induction has mostly worked in the past to justify induction is not viciously circular since it uses induction but does not assume that induction is justified. Still, we grant that this inductive justification of induction is disappointing and does not compare well to Quine's use of evolutionary theory to justify induction. Why is it disappointing in comparison with Quine's appeal to evolution? There are two grounds for disappointment. The first is that this inductive basis for justification is too vague to be robustly testable and hence while it is not entirely vacuous and not viciously circular, the justification is comparatively lacking in substance. Suppose induction fails on occasion to predict correctly. Do we reject our use of induction as without justification? Of course not. Suppose it fails a hundred times or a thousand or ten thousand or more. When do we know that induction has not mostly worked in the past?

Though similar charges of untestability have been leveled against evolutionary theory, especially against Darwin's principle of natural selection that forms part of it, such worries no longer have much force among scientists. Indeed, the shape of evolutionary theory is today constantly being updated because aspects of the theory in the context of other assumptions are subject to robust tests against the evidence (Kitcher 2007; Campbell & Robert 2005). Moreover, if we shift our critical gaze from induction in the abstract to the cognitive mechanisms underlying this capacity there is a large amount of empirical and normative work to do. The recent interest in implicit bias in philosophical circles (Brownstein 2015) offers one example of where a specific kind of untutored inductive mechanism has come under critical scrutiny and philosophers are at pains to provide tools that will ground better inferences than these biased psychological tendencies.[13] If we are interested in better understanding the success rate of the various mechanisms underlying the basic inductive tendency we can look to comparative psychology and ethology to empirically investigate.

The second ground of disappointment is equally important and connected with the first. If we believe that induction has mostly worked in the past we have some (inductive) justification for thinking it will work in the future, but there is enormous room for doubt at the same time if we don't have any idea why the world we live in and our brains that interact with it are so formed that we have better than coin tossing chances of getting it right when we follow our inductive instincts. The theory of evolution offers precisely such an explanation. Unlike the simplistic appeal to what has worked in the past, the evolutionary justification is manifestly not simplistic in providing a (testable) *explanation* of why induction has worked in the past. Though both kinds of justification fail to be viciously circular, only the appeal to evolutionary theory is robustly testable and powerfully explanatory.

The proposed justification faces a second problem, of incompleteness, since use of 'our innate, scientifically unjustified similarity standard,' to use Quine's words, is obviously only part of the contemporary practices of induction. There is, for instance, also the matter of justifying scientific theories that can be used to modify or in some cases reject the untutored inferences that we make about how similar the future will be to the past in specific respects. Therefore, even apart from the problem of normativity, the Darwinian story cannot hope to justify induction *tout court*. The appropriate response is twofold. First, concede that the account of induction is incomplete, as Quine makes amply clear in his statement of the solution when he says, 'Darwin's explanation is a plausible *partial* explanation [of why our inductions do better than chance]' (Quine, 1969, 127, emphasis added). Second, argue that justifying the basic inductive tendency is no less significant for being partial, since it plays a *foundational* role in induction as a whole. The reason is simple and already stated: without the primitive inductive tendency our ancestors would not have survived to reproduce their kind and there would be no inductive practice to justify in whole or part. Obviously this tendency is not foundational in the sense of being infallible. As just noted, it can lead to error and often science, based on induction applied to theory, can explain by virtue of theory why and

---

[13] Note, there is no reason to think the implicit biases that have been the target of this analysis are untutored or evolutionarily ancient.

how the innate tendency errs. Still, it does better than chance, and without it we could not build a more advanced, general understanding of induction in science. Imagine if our everyday world were totally unpredictable.

Before leaving this topic, it is worth noting a different charge of incompleteness: that the evolutionary explanation by its nature concerns the past, not the future, and thus misses the classic problem, which is, as Quine says, about whether we are justified in believing that our *predictions* based on induction will come out right more often than chance. The explanation given, of course, is not simply about why our inductive tendency came to exist in the first place, but about why it continues to exist today and can reliably inform us about the future. But then the theory implies that, unless the environment radically changes so that having the tendency is no longer advantageous, predictions based on it will continue to be more reliable than chance. The theory like other scientific theories is, after all, a basis for predictions about the future. Though it is theoretically possible that a theory in the future may be inductively supported that predicts our basic inductive tendencies will begin to do worse than chance and hence cease to be justified, we have at present no inductive grounds to believe this dire prognosis and certainly none in evolutionary theory. Current evolutionary theory, in sum, supports the continuing rough reliability of our evolutionarily ancient inductive tendency.

We can see, then, that objections from normativity, incompleteness, and circularity are unable to undermine the appeal made by Quine to evolutionary theory to both explain and justify induction. It does not follow, needless to say, that such an appeal must be beyond criticism. A range of other kinds of objection are no doubt worth consideration that address the scientific claims that are implied or presupposed in the appeal to evolutionary theory, for example, the basic claim that our natural tendency to make inductive inferences evolved by natural selection rather than evolving through some other mechanism. We set those concerns aside. Our object to this point has been only to refute objections aimed at the very idea of justifying our basic inductive tendency by evolutionary theory, as if the endeavor should fail whatever the details of the science. Such details matter a lot. The most serious lacuna produced by our discussion so far is the incompleteness of our account—missing, as it does, many of our contemporary inductive practices. However, this is exactly where we find the Logical Project offers a particularly powerful explanatory story, so we now turn to Kitcher to help us complete the picture.

3.3 The Logical Project

Kitcher understands ethics differently from the way it is commonly viewed by those who believe that ethical knowledge is possible, namely as knowledge based on principles of right and wrong that are external to us and beyond our control, such as principles divinely revealed or discovered by brilliant thinkers. Instead, on his pragmatic and naturalistic understanding of ethics, human ethical knowledge and practice are part of a long cultural evolutionary process in which humans tried to work out among themselves how they can live together and prosper by imposing constraints on their interactions and setting goals to strive toward collectively. Their first efforts emerged out of primitive tendencies toward ethical thinking fashioned through natural selection much as our early efforts to understand the natural world emerged from primitive inductive tendencies. The 'Ethical Project' as he calls it is on-going with no final, finished set of principles to live by, just as our later efforts to comprehend how best to make inferences is also on-going and constantly under revision. Our Logical Project, as we think of it, is parallel in its open-endedness and in its origins to Kitcher's Ethical Project. Retooling Kitcher, 'The [logical] project evolves indefinitely. Progress is made not by discovering something independent of us and our societies, but by fulfilling the functions of [logic] as they have so far emerged' (2011, 285-6).[14]

So far we have an account of the evolution of inference that parallels the first part of Kitcher's analytic history in *The Ethical Project* (2011, chapters 1-2). Kitcher argues that for cognitively complex social organisms altruism will likely evolve and, in much the same vein, Quine shows that for cognitively complex organisms basic induction will likely evolve. We thus get a natural historical starting point for the capacity to make good inferences. Kitcher suggests that somewhere in the hominid line a new kind of cultural evolution began shaping the expression of our evolved altruistic tendencies and it is this that marks the real beginning of the Ethical Project. Although we would dispute Kitcher's anthropocentric assumption that this cultural evolutionary process could only have started in hominids, the basic idea that cultural evolutionary processes have been the main drivers of the Ethical Project, as well as the Logical Project, seems to us importantly right. For both projects cultural evolution is the main shaping force, determining the histories of ethics and logic leading up to the specific theories and practices that we see today. Biological evolution can generate basic cognitive capacities that produce better than chance inferences and even the ability to reflect on those inferences, but norms for evaluation and practices of deliberation can only be produced in a social context and survive into the next generation through culture.

What is it that starts this new cultural trajectory? In *The Ethical Project* Kitcher explains how the conflict between altruistic tendencies (evolved through group selection) and individualist, selfish tendencies forms a problem space where cognitively complex hominids with sophisticated communication strategies could deliberate and come to agreements as to how to solve these conflicts and ultimately live better. When these agreements were successful they allowed the development of new forms of social life, which then provided new opportunities for altruism, selfishness, deliberation and problem-solving thus creating a feedback loop that, at least on the grand scale, has tended to be progressive, or so Kitcher argues (2011, 285-329). The Logical Project does not need quite such a rarified problem space for cultural evolutionary process to get going, though like the Ethical Project it is fundamentally about problem solving—inferential problems rather than ethical ones. All that is required is for a being to recognize that her or his inferences (e.g., Quine's evolved inductive tendencies) are

---

[14] This conception of the Logical Project can be viewed as compatible with justifying changes in logic by appeal to the Goodman-Rawls-Daniels notion of Wide Reflective Equilibrium. That in turn suggests that other disciplines, such as psychology, could play a role in shaping the direction of logic. Some philosophers (e.g., Resnik 1985) see other subjects as irrelevant to logic as a purely normative discipline. We address in section 3.4 the idea of logic 'as it is in itself' apart from how it is used or taught.

3.3. THE LOGICAL PROJECT

not perfect in addition to the desire and ability to make them better. Of course, this involves a fair degree of cognitive complexity. First, it requires higher order reflection on action or judgement. Second, in order for improvements to inferential practices to spread throughout a population and be passed from generation to generation, there must be a means of transmitting this information. Thus the Logical Project is a social project where communication and deliberation are particularly powerful engines, producing effective, distinctive solutions to inferential problems and allowing us to revise these solutions should we come up with something better. Some inferential solutions make possible new and unanticipated ways of living—consider, for example, the extraordinary power of probability theory and its subsequent use in social statistics and policy making. Thus we see logic as a kind of social technology which produces new problems just as it solves others.[15]

Here we find the idea of a cultural ratchet particularly useful. For the present discussion the cognitive capacities that underlie such ratchets are not of central interest, though there is a nascent, contentious literature on the topic.[16] Instead we are interested in the aetiological structure of ratchets. In the cultural transmission of traits or behaviours, the 'ratchet effect' is where 'modification and improvements stay in the population fairly readily (with relatively little loss or backward slippage) until further changes ratchet things up again' (Tennie, Call and Tomasello 2009, 2405). Thus individuals within a population are spared the need to rediscover the same insights generation after generation, but can instead build from the hard won lessons of their ancestors. Cooperation and, more specifically, teaching are perhaps the primary mechanisms for transmission, allowing for the accumulation and modification of cultural practices (Tennie, Call and Tomasello 2009, 2411).

We can now see why language is likely to have been an extremely powerful ratchet in the Logical Project. This is most obvious with deduction. Insofar as the formal logical project is crucially about syntactic structures that strictly preserve truth, it can only emerge once there is an abstract, symbolic representational system like natural language.[17] Whatever the natural historical origins of language itself, for a formal logical project with strict rules of entailment operating on non-abstract content to get off the ground our ancestors presumably required symbolic forms of representation (i.e., sentences) that could be understood as true simpliciter, and not subject to the changeability of environment and interests that accompanies unreflective experience. The idea here is not that thought or inference is impossible without language; both Bermúdez and John Pilley's work suggests that at least some nonhuman animals are capable of even deductive reasoning (Bermúdez 2007, 139-149; Pilley and Reid 2011,

---

[15] Interestingly, the Ethical Project appears to depend on the Logical Project and indeed, the only authority that Kitcher cedes to philosophers in the Ethical Project is as experts in deliberation, providing insight into methods of argumentation and good inference.

[16] Much of the literature looking at the cognitive basis for cultural ratchets, including Tennie, Call and Tomasello (2009), which we follow here, treats cultural ratchets as a means for explaining human uniqueness. This displays a question-begging anthropocentrism that we reject. It is, of course, possible that we might *discover* that cultural ratchets are uniquely human but *defining* the term as such seems more an expression of political ideology than a credible application of scientific method.

[17] We say 'like natural language' here in deference to the existence of deductive entailments in practices that are not linguistic, such as geometry.

193). Rather, the point is that language tends to shape thought into comparatively decisive, clear, and robust content that allows more complex formulations and strict preservation through various inferential processes.

While it is easier to imagine how nonlinguistic animals, hominids or otherwise, might develop culturally transmitted informal inferential norms—for instance, an appeal to authority taking the form of deference to more dominant or elder members of a social group—the kind of sophisticated inductive practices that characterize modern science would be impossible without natural and mathematical languages. We can still see traces of our evolutionarily ancient, untutored inductive tendencies in the unreflective judgements of similarity that make scientific observations possible and what might be called scientific intuition or hunches (see Eve Roberts' account of hunch hypotheses [Roberts 2014, §4.4]). However, given how central enculturation is to any human developmental path that might be considered normal or healthy, it would be naïve to suppose that these species-typical characteristics are not shaped by a variety of cultural norms, at least in part. Another way of putting this is that for any typical, mature member of a highly social species that exhibits rich and diverse cultures, no inference is likely to be entirely untutored or innate.

What is missing so far are any of the details of the cultural evolutionary path that takes us from our Quinean origins to the present. But that path is the entire global history of logic. Even if we limit our vision to a Whiggish history of the philosophy of the so-called West, this will take us through a huge amount of deliberation, revision, transmission, and ratcheting. The formal story would take us from Aristotle's syllogism, through the many innovations of the Middle Ages, through the mathematicization of logic in the 19th century to the logical pluralism that exists today. The informal story, even if just restricted to the sciences, would take us from any ancient or medieval discussion of argumentation, through Baconian method, through hypothetico-deductivism and probability theory, to the many sophisticated methods of the contemporary sciences. A cursory knowledge of these traditions reveals the tell-tale signs of the ratchet effect. New theories and applications spring from old ones, not from whole cloth, and they retain some of the ideas and practices from their predecessors even as they reject others. The increased logical power of the new theories and the benefits of logical innovations for other cultural practices often make it difficult to simply go back.

Despite the esoteric, even monkish, reputation of logic, we can see how various innovations have radically changed human cultures. Computational theory and electronic computers are cases in point, as it is hard to overestimate the extraordinary effects they have had on contemporary life around the globe. Moreover, the computing power of even ordinary devices, like the laptops on which this paper was written, has had an astonishing impact on our current inferential processes making enormous data sets tractable, performing mathematical proofs, and allowing simulations, just to name a few recent innovations that have radically changed the natural and social sciences (see Eve Roberts 2014).

3.4 What is Logic? Revision

Thinking of the history of logic as a cultural evolutionary process challenges us to reflect on what exactly logic is. The Logical Project treats it as normative and evolving—as continually under revision—which will no doubt sound like heresy to many. Indeed, to suggest that logic is changeable may seem like a category mistake. But what should we infer from the huge number of revisions that we see in the history of logic? How can we know that these changes are good ones and that the practices and theories that we currently have are the best? We find a problem of normativity that is similar to the one that we discovered when we considered our evolutionarily ancient untutored inductive tendency in section 3.2, above. The Logical Project can describe how logic has evolved, albeit through cultural selection pressures rather than strictly biological ones, but what is the justification for thinking that the resulting inferential practices are rational and that we should follow them?

Happily, Graham Priest has recently addressed this question in 'Revising Logic' (2014) and we find that much of his analysis fits neatly into the Logical Project. Although Priest remarks that much the same approach might be applied to non-deductive logic (211), he narrows his discussion to deduction alone. He begins by offering a useful analysis of the several different things that we might mean when we ask whether logic can be rationally revised. Employing concepts of the medieval logicians, he distinguishes '*logica docens* (the logic that is taught),' which refers to the kind of thing we find in logic textbooks, from '*logica utens* (the logic which is used)' (Priest 2014, 212). Priest believes that this distinction is not exhaustive so he coins a term of his own, '*logica ens* (logic itself)...[which] is what *is* actually valid: what really follows from what' (Priest 2014, 212), emphasis his). With these distinctions in hand, Priest characterizes the history of logic in such a way as to suggest that it will always be undergoing revision, i.e., evolving. Moreover, although he doesn't come out and say as much, the account he offers leaves no real role for *logica ens* and pegs the rational revisability of logic on the construction of systems that allow us to make ever better inferences in the many applications in which we employ them. For Priest, logical theories and practices are tested in much same way that scientific theories are tested—against themselves (an internal consistency condition) and against the world. Thus we will find that we have an instrumental justification for trusting our current logical systems and practices that looks very much like the Quinean version, discussed above.

As Priest points out, it is clear that from Aristotle's Lyceum, to the medieval universities and monasteries of the Middle East and Europe, to university classrooms today, we have seen many revisions in *logica docens* (212-3). The trickier question is whether these revisions were rational. Here Priest draws an interesting analogy with logic's sister science, mathematics. He points out that from a mathematical perspective the many different pure geometries that we have—Euclidean, Lobachevskian, Riemanian, and so forth—are all equally good. It is only in application that we can find rational grounds for preferring one over another. 'Each applied geometry becomes, in effect, a theory of the way in which the subject of the geometry behaves' (215). Now there is a kind of canonical application, which is implicitly assumed when no other more specific application is in force. For geometry this is 'the spatiotemporal structure of the physical cosmos' (215). In this application it is a kind of discovery that the correct geometry is non-Euclidean, even though from the perspective of pure mathematics, Euclidean geometry is just as good as any other. Similarly, Priest argues, we have any number of different pure logics but which one is right depends on the application in question. For instance, and we will just take Priest's word for this, classical propositional logic is the logic best suited to simplifying circuits, while the Lambeck calculus ably parses grammatical structures (215). As with geometry, there is a canonical application for logic when no more specific application is in force. It is hardly surprising that this conforms to the function of logic outlined in section 3.2.1, above. The canonical application of logic is the ordinary practice of reasoning. Thus 'a pure logic with its canonical application is a theory of the validity of ordinary arguments: what follows (deductively) from what' (216).[18] The rationality of revising a logical theory depends on judging it in its application.

Fascinatingly, Priest identifies the rationality of the revision of logical theories as resting on the same norms as the revision of any other theory. However the norms he lists are those most centrally associated with the natural sciences: adequacy to the data, simplicity, unifying power, non-ad hocness, fruitfulness, and so forth (217). Of course, as Kuhn pointed out when he clearly articulated these values (1977), there is no single rational ordering of them and thus decisive answers as to which of a set of competing theories is best and why may often be lacking. Nonetheless, they are typically robust enough to determine, in Priest's words, 'the [logical theory] that comes out best on balance' (217).

Priest's analysis of *logica utens* is largely parasitic on this account of *logica docens*. It is important not to mistake *logica utens* for a psychological notion about how we reason as it is clear that human reasoning is systematically prone to certain types of error. (These errors range from our tendencies to affirm the consequent (Wason 1968), to our tendencies to make systematic errors in probabilistic reasoning (Tversky and Kahneman 1986), to pervasive discriminatory implicit biases (Brownstein 2015)). Instead, *logica utens* is a normative notion picking out the norms of an inferential practice (Priest 2014, 218). Here again we have the question of rational revision, but Priest solves this deftly by simply requiring that to be rational we must bring our *logica utens* in line with our best current *logica docens* (219). As Priest notes, 'How else could one be rational about the matter?' (220).

There are a few things that strike us about Priest's account of rational revision. First, these empirical values are themselves a product of the inductive branch of the Logical Project and historically these inductive norms developed parallel to the deductive systems to which Priest applies them. In this light, Priest's remark that an analysis similar to the one he offers might be given for induction seems somewhat glib. In fact, Priest gives induction a foundational role in the justification of *logica docens*. Now, he might object that this is the only conceivable way that we can test deductive logic. Quite so, but then we simply have to admit to the intimate relationship between

---

[18] Nothing in Priest's account precludes the development of formal systems without any application, but simply for the fascination of doing so—a truly pure science endeavour. Here again the history of geometry is instructive. After all, initially the development of non-Euclidean geometries in the 19th century was an exercise in pure mathematics, until people like Einstein put them to physical use.

3.4. WHAT IS LOGIC? REVISION

inductive and deductive logic, when considered on a grand historical scale. Second, Priest's goal for the revision of logical theory—'coming out best on balance'—sounds a good deal like Quine's goal for the basic inductive capacity—'coming out right'. The cultural path of the Logical Project marches toward the same goal as the natural selective processes that underlie it—having a better and better chance of getting things right when we make inferences about how the world is or will be. Just as the world, as discovered and described through Einsteinian physics, pushed back on geometry to replace Euclidean geometry in its canonical application, so the relevant parts of the world—be they biological, physical or technological products—push back on different logics in their varied applications. If Priest is right, choices of a logic are pregnant with ontological assumptions about the part of the world to which we apply it. As our empirical and logical projects evolve, logical insights can drive us to revise our understanding of the world and, likewise, empirical insights can drive us to revise our logic.

Finally, Priest considers *logica ens*—'the truths of the form "that so and so follows from that such and such"' (220). Initially, one expects this notion to be a strictly realist idea, but Priest warns us off such a reading. He writes, '...logic is not a natural science. It is a social science, and concerns human practices and cognition. When a theory changes in the social sciences, the object of science may change as well,' (220) though he is quick to note such change is not inevitable. The theory of *logica ens* is fundamentally a theory of validity, so the question of whether *logica ens* is revisable is a question of whether validity is revisable. Of course, whether validity is revisable or not depends on what validity actually is. But as he points out, identifying what validity is means determining our best theory of validity. Now, even if we suppose that 'true' validity inheres in some eternal Platonic realm of abstract objects—such as the sets of a model-theoretic approach or the proof-structures of a proof theoretic account (221)—our expression of these truths does not share in this permanence. We describe these abstract objects and their theories in natural language; meanings in natural languages change as our applications and uses of terms change; and thus the sentences expressing validity claims change their truth values. This suggests that even if someone were to express and justify a final *logica ens* truth on some matter of validity, due to rational processes of meaning revision (an inevitable effect of *logica utens*), the sentence expressing this matter would not remain eternally true, and the error would eventually come clear in our *logica docens*. Although Priest seems committed to some role for *logica ens*, given the feedback loop between *logica utens* and *logica docens*, we find it difficult to see how it could have any independent influence on either logical theory or logical practice.

Unless we can find some nonmiraculous method of connecting *logica ens* to logical practice it seems we must relegate it to assessing the consistency of internal structure of formal systems. If we suppose that *logica ens* is revisable, then it seems it must be conventional. But if it is conventional then it is determined by the practices of *logica utens* and *logica docens*. If it is not revisable it is difficult to see how we could ever come to know it. After all, if we compare the set of all postulated necessary, analytic or *a priori* truths against the set of those that Platonically-inclined philosophers would endorse today, we find something like Laudan's pessimistic meta-induction (1981), where it seems absurdly arrogant to suppose that while everyone in the past was wrong we, right here, right now, have got it right. Quine's (1951) argument against the concept of analyticity is further and even more compelling grounds, for rejecting the idea that there are abstract universal truths about validity itself that sit independent from and prior to our logical theories and practices of inferring. When we actually do the work of figuring out the inferential structure of formal systems we are doing the work of *logica docens* and *logica ens* thus evaporates.

Here we find a remarkable resonance between Priest's concluding remarks on the revisability of logic and our Kitcher-inspired Logical Project. Priest summarizes his position:

> The rational *logica utens* depends on the rational *logica docens*. The true *logica docens* depends on the facts of validity. And assuming a model- or proof-theoretic account of meaning, the language available to express these may depend on the *logica utens*. It is clear that we have a circle. If one were a foundationalist of some kind, one might see this circle as vicious: there is no privileged point where one can ground the entire enterprise, and from which one can build up everything else. However, I take it that all knowledge, about logic, as much as anything else, is situated... We are not, and could never be, *tabulae rasae*. We can start only from where we are. Rational revision of all kinds then has to proceed by an incremental and possibly (Hegel notwithstanding) never-ending process. (223)

The Logical Project puts this incremental and possibly never-ending process into a larger famously incremental and never-ending process—evolution. To retool Kitcher's words, '[Logic] emerges as a human phenomenon, permanently unfinished. We, collectively, made it up, and have developed, refined, and distorted it, generation by generation. [Logic] should be understood as a project—the [logical] project—in which we have been engaged for [all] of our history as a species' (Kitcher 2011, 2).

3.5 Conclusion

The account we have given shows how the methodological core of philosophy might be thought of as the product of biological and cultural evolution. Importantly, this makes logic an on-going project aimed at making good inferences and constructing formal systems rather than a set of fixed truths waiting to be discovered. Considered from this perspective we can liberate logic as a discipline, in both quotidian and expert practices, from supernaturalist appeals to things like Platonic realms, without sacrificing a robust normativity. The value of logical systems is grounded in the fact that we value making good inferences. Whatever the natural history of this evaluative fact, through our cultural history we have built an astonishing array of extremely powerful and reliable logical systems from this starting point, through which we have transformed our world. A naturalized approach like ours reframes current innovations in logic, such as paraconsistent logics, as new ways of responding to inferential challenges that may be useful additions to current systems rather than replacements for them. How we make revisions to these systems, applying rules of consistency and evidence often external to the systems themselves, is a matter that is itself up for revision—a coevolving metalogical project. Even as we acknowledge the role for innovation, in its most abstract formulation, the inferential problems of our modern world remain the same as those of our most distant ancestors. It is the problem named by Quine: how can we, 'as we now are..., in a world we never made,... stand a better than random or coin-tossing chances of coming out right'?

Thanks We would like to thank Austin Booth, Tyler Brunet, Ford Doolittle, Andrew Fenton, Jobin Kanjirakkat, Carlos Mariscal, and Gillman Payette for their spirited and constructive comments, in conversation and in writing, occasioning many needed revisions. Thanks especially to Gillman Payette for his tireless editorial support.

3.6 Bibliography

Bermúdez, J. L. (2003). *Thinking Without Words*. Oxford University Press.

Braybrooke, D., B. Brown and P. K. Schotch (1995). *Logic on the Track of Social Change*. Clarendon Press.

Brownstein, M. (2015). Implicit Bias. *The Stanford Encyclopedia of Philosophy* (Spring 2015 Edition), Edward N. Zalta (ed.), <http://plato.stanford.edu/archives/spr2015/entries/implicit-bias/>.

Call, J. (2006). Descartes' two errors: Reason and reflection in the great apes. In S. L. Hurley and M. Nudds (Eds.), *Rational Animals?* Oxford University Press.

Campbell, R. (1998). *Illusions of Paradox: A Feminist Epistemology Naturalized*. Rowman & Littlefield Publishers.

Campbell, R. and J. S. Robert (2005). The structure of evolution by natural selection. *Biology and Philosophy 20*(4), 673–696.

Gillon, B. (2011). Implicit Bias. *The Stanford Encyclopedia of Philosophy* (Summer 2011 Edition), Edward N. Zalta (ed.), URL=<http://plato.stanford.edu/archives/sum2011/entries/logic-india/>.

Elgin, C. Z. (2004). True enough. *Philosophical Issues 14*(1), 113–131.

Gendler, T. S. (2008). Alief and belief. *Journal of Philosophy 105*(10), 634–663.

Goodman, N. (1955). *Fact, Fiction & Forecast*. University of London.

Greenwald, A., D. McGhee, and J. Schwartz (1998). Measuring individual differences in implicit cognition: The implicit association test. *Journal of Personality and Social Psychology*, 74: 1464-1480.

Hurley, S. L. and M. Nudds (2006). *Rational Animals?* Oxford University Press.

Kitcher, P. (1993). Function and design. *Midwest Studies in Philosophy 18*(1), 379–397.

Kitcher, P. (2007). *Living with Darwin: Evolution, Design, and the Future of Faith.* Oxford University Press.

Kitcher, P. (2011). *The Ethical Project.* Harvard University Press.

Kitcher, P. (2012). Précis of the ethical project. *Analyse & Kritik 34*(1), 1–19.

Kornblith, H. (1993). *Inductive Inference and its Natural Ground.* MIT Press.

Kuhn, T. S. (1977). Objectivity, value judgment, and theory choice. In *The Essential Tension*, pp. 320–39. University of Chicago Press.

Laudan, L. (1981). A confutation of convergent realism. *Philosophy of Science 48*(1), 19–49.

Longino, H. (1994). The fate of knowledge in social theories of science. In F. F. Schmitt (Ed.), *Socializing Epistemology: The Social Dimensions of Knowledge*, pp. 135–158. Rowman and Littlefield.

Pilley, J. and A. Reid (2011). Border collie comprehends object names as verbal referents. *Behavioural Processes 86*(2), 184–195.

Priest, G. (2014). Revising logic. In P. Rush (Ed.), *The Metaphysics of Logic*, pp. 211–23. University of Cambridge Press.

Quine, W. V. O. (1951). Two dogmas of empiricism. *Philosophical Review 60*(1), 20–43.

Quine, W. V. O. (1969). Natural kinds. In J. Kim and E. Sosa (Eds.), *Ontological Relativity and Other Essays*, pp. 114–38. Columbia University Press.

Quine, W. V. O. (1981). *Theories and Things.* Harvard University Press.

Resnik, M. D. (1985). Logic: Normative or descriptive? The ethics of belief or a branch of psychology? *Philosophy of Science 52*(2), 221–238.

Roberts, E. (2014). *System-driven research: Legitimate experimental design for biological/biomedical research.* Phd., Dalhousie University, Halifax, Nova Scotia.

Schechter, J. (2013). Could evolution explain our reliability about logic? In T. S. Gendler and J. Hawthorne (Eds.), *Oxford Studies in Epistemology 4*, pp. 214-39.

Schotch, P. K. (Forthcoming) *Essays in Philosophical Logic*

Schotch, P. K., Brown, B. and Jennings, R. E. (Eds.) *On Preserving: Essays on Preservationism and Paraconsistent Logic* University of Toronto Press, Toronto, 2009.

3.6. BIBLIOGRAPHY

Tennie, C., J. Call, and M. Tomasello (2009). Ratcheting up the ratchet: On the evolution of cumulative culture. *Philosophical Transactions of the Royal Society B: Biological Sciences* (364), 2405–15.

Tversky, A. and D. Kahneman (1986). Rational choice and the framing of decisions. *Journal of Business 59(4)*: S251-78.

Wason, P. (1968). Reasoning about a rule. *Quarterly Journal of Experimental Psychology 20*, 273–81.

Chapter 4

Identity, Haecceity, and the Godzilla Problem

Kent A. Peacock[1] and Andrew Tedder[2]

4.1 Introduction: An Embarrassing Problem

If one's only knowledge of logic came from standard university texts, one might think that elementary first order predicate logic with identity has all been worked out a long time ago and that there are thus no serious technical or conceptual problems lurking within it. We're not sure that this comfortable view is right.

We're going to begin by pointing to what we believe to be an embarrassing problem for standard first-order predicate logic with identity. The usual approach is to take the self-identity of all objects in the universe of discourse as a logical truth or theorem; that is, it is taken that

$$\vdash \forall x (x = x); \tag{4.1}$$

or as

$$\vdash a = a \tag{4.2}$$

in a scheme allowing generalization to (4.1). The symbol $\vdash$ in these formulas is not meant to suggest that they are provable from the other resources of first-order logic; but rather that they are assertions that may be added without proof to first order predicate logic to give a theory of identity. In Lemmon's system (1978) the rule (4.1) is called Identity Introduction and symbolized =I. It is paired with Identity Elimination (=E), which says that if $a = b$ then a may be substituted for b or *vice versa* wherever they occur. Identity Elimination is the motor that drives virtually all of all applications of identity. Identity Intro, however, is rarely used, and as the following natural deduction proof shows, it allows consequences with which all logicians ought to be uncomfortable:

---

[1] Department of Philosophy, University of Lethbridge, kent.peacock@uleth.ca
[2] Department of Philosophy, University of Connecticut, andrew.tedder@uconn.edu

(1) $\forall x(x = x)$ =I
(2) Godzilla = Godzilla 1 UE
(3) $\exists x(x = \text{Godzilla})$ 2 EI

So we have

$$\vdash \exists x(x = \text{Godzilla}). \tag{4.3}$$

We'll call this argument pattern the *Categorical Godzilla*. (There is also a *Conditional Godzilla* which we will later introduce.) If $\forall x(x = x)$ is a logical truth, then we apparently have it as a *logical truth* that a certain city-trampling movie monster really does exist. This pattern could be repeated for any name whatsoever. *What is wrong with this picture?*

Of course, this problem is well known (or should be) and it is one of the motivations for the construction of various types of free logic, the defining characteristic of which is that names are not automatically assumed to refer (Lambert, 1991; Bell et al., 2001; Nolt, 2011). We are sympathetic to free logic and we think that our observations here tend to give further motivation for its development. However, our primary aim in this note is the more limited goal of arguing against the theoremhood of =I. In doing so, we will show that there is reason to question =I even as a universal assumption for at least some of the domains to which predicate logic might be applied. We will also show that a consequence of treating =I as a global but defeasible assumption is, surprisngly perhaps, the resurrection of Carnap's *null thing* (1947), or something very much like it.

4.2 A Short History of Arguments for =I

Many currently used logic texts get around the awkward conclusion that =I can be used to prove that anything whatsoever exists simply by not mentioning it at all. E. J. Lemmon's widely used *Beginning Logic* (1978) relegates it without comment to an exercise, and defends the theoremhood of =I as follows:

> For any term t, the rule =I permits us to introduce into a proof at any stage $t = t$, resting on no assumptions. The idea should be clear: anything is itself, as a matter of logic; hence $t = t$ is logically true, and so can appear without assumptions. (Lemmon, 1978, p. 161)

It is by no means clear that everything being itself is a matter of logic. By contrast, the theorem $P \to P$ is a matter of logic, and indeed it is sometimes (confusingly) called Identity; it is almost as if Lemmon confused = with the truth-functional connective $\to$, or perhaps $\equiv$.

A recent logic text by Paul Herrick trades on intuitions similar to Lemmon's:

> Each thing is identical with itself ... Surely this needs no argument; certainly it is necessarily true. (How could something possibly *not* be identical to itself?) (Herrick, 2013, p. 587)

One should be suspicious of arguments for p of the form, "Surely p...". For surely (if we may) Lemmon and Herrick are appealing not to logical intuitions about identity, but

4.2. A SHORT HISTORY OF ARGUMENTS FOR =I

metaphysical intuitions about identity. They are saying that it is a matter of *necessary fact* that every item is self-identical, and they have forgotten that pure logic as such expresses no facts. Facts, whether necessary or of the ordinary garden-variety, can be introduced into a logical problem only *by assumption*. The views of Lemmon and Herrick therefore seem to be part of a long tradition of mistaking presumed factual or metaphysical necessity for logical or mathematical necessity.

In *Principia Mathematica* Russell and Whitehead (1927, *13) gave an apparently much more principled and precise defence of =I. They begin by defining = by means of the Identity of Indiscernibles: $x = y$ means that if ϕ is a property of x then ϕ is a property of y. (In this sketch of their exposition we gloss over niceties having to do with the Theory of Types.) The definition gets turned into a theorem by an application of universal instantiation: since it is true of any two arbitrarily selected entities that they are identical if and only if they share all properties, then that holds for all instances of $x = y$. Then as a special case of this result any arbitrarily selected entity is self-identical simply because any property of itself is a property of itself.

The Russell-Whitehead approach has the virtue of precision. However, modern authors tend to shy away from it because it requires second order logic. More important from the skeptical point of view we pursue here, the reliance on the Identity of Indiscernibles again amounts to building a metaphysical principle into formal logic. It could even be said (though no doubt contentiously) that Russell and Whitehead's view, that all terms are self-identical because any arbitrary term would have the same properties as itself, borders on question-begging. Russell and Whitehead do not state the Categorical Godzilla, though it would be readily available in their system.

Reliance on the Identity of Indiscernibles in order to define identity and justify theorems about it can be traced back through Frege (1967) and Leibniz (discussed in Kneale and Kneale (1962)) to Aristotle. The latter seems to be the first to have introduced the concept of the Identity of Indiscernibles, though informally, in *De Sophisticis Elenchis* (Aristotle, 2012, Ch. 24 (179a37)): "For only to things that are indistinguishable and one in essence is it generally agreed that all the same attributes belong." His wording "it is generally agreed" suggests that this principle has a longer history, either orally or in writings now lost. Being "one in essence" is a metaphysical requirement for self-identity; Aristotle's discussion surrounding the line quoted here explains why identity may otherwise be ambiguous if this high metaphysical standard is not met.

The reliance upon the Identity of Indiscernibles to get =I is therefore very old. Now =I, for our purposes, can be treated as either substantive or simply stipulative. If the latter, then it's the kind of thing that we may choose to do without. If the former, then its place in the reasoning designed to justify logic is dubious at best. Logic should not be a substantive inquiry, but rather a formal one (this is how the field has been moving, and we think, for the better). What we should be interested in are extremely generalized relations between assertions about objects burdened with as few assumptions as possible. So how can we amend standard first order logic with identity in as conservative a way as possible, but so as to block the Categorical Godzilla?

4.3 Blocking the Godzilla Inference

There is no way to block the Godzilla proof by placing some sort of artificial restriction on EI without crippling or drastically reconstruing EI, which is not in line with our conservative approach. And the application of UE to line (1) seems to be entirely unobjectionable. One could simply introduce an *ad hoc* rule against making inferences of the Godzilla type; for instance, one might introduce a rule against applying EI to any formula of the form $a = a$. This probably would be logically possible but it seems inelegant. To begin with, we want to change as little as possible of classical first order logic, and only make such changes to identity theory as would be sufficient to block Godzilla in a natural way. (Further along we'll suggest something more radical.) It is therefore much more pertinent to examine whether it really is reasonable to treat $\forall x(x = x)$ as a theorem.

The first point to note is that the Godzilla proof is simply an illustration of the rule that there is an existential claim built into any proposition of the form Fa. Suppose we assume or are given Fa, where F is some property and a is an individual. From this we conclude $\exists x Fx$ by an immediate application of EI. So to assert that an individual does in fact possess a property is to imply the existence of an individual possessing that property, and this is true for any property F, *including self-identity*.

This suggests that another way of blocking the Categorical Godzilla might be to question whether self-identity really can be treated as if it were monadic property. If Godzilla is a city-trampler then something is a city-trampler; if Godzilla is self-identical then can we say that something is self-identical? It would be very odd if we could not, since self-identity certainly *is* something that pertains to individual entities when it pertains at all; hence, that approach does not seem promising either.

The key is that we do not want to build any assumptions about the existence of any entities whatsoever into our logic. Therefore, we can avoid the Categorical Godzilla if (i) we remind ourselves that $\forall x Fx$ and Fa can be introduced in a proof only as *assumptions* and (ii) insist that this rule be followed even when F is self-identity—and even if we find it hard to *imagine* that any given entity could not be self-identical. On some metaphysical views it might indeed be the case that everything is self-identical, but this can't be a matter of *logic* even if it is a necessary truth, whatever *that* might be. There are certainly some universes of discourse that contain only self-identical objects (such as the universe comprised of the set of natural numbers), but our choice of a universe of discourse is not a matter of logic either.

A possible defence of the orthodox view could be along the following lines: it could be said that in doing first-order logic we always take it for granted that the universe of discourse $\mathcal{U}$ consists of objects that are already presumed to exist. We see two objections to this view.

First, even if we want to say that we take it from the outset that all items in $\mathcal{U}$ exist, we do not mean to say that it turns out to be a *theorem* that some item in $\mathcal{U}$, which might be, say, the city of Paris, France, happens to exist; rather, the existence of any item in the actual world is an empirical matter if we are talking about real-world entities (unless we could view the world from the perspective of the God of Leibniz for whom all apparently empirical truths are analytic), or a mathematical matter if we are talking

4.3. BLOCKING THE GODZILLA INFERENCE   67

about mathematical entities such as sets or numbers.

Second, logic would not be very useful if we only allowed ourselves to talk about things that we already know or believe to exist. One of the most powerful tools of thought is the ability to consider hypothetical objects without existential commitment. This is just what we do in indirect proof in mathematics. Consider Euclid's proof that there is no greatest prime: it begins by supposing hypothetically that there is a greatest prime, it gives this hypothetical number a name for convenience in calculation, and then shows that such a number must have contradictory properties. The universe of discourse should be precisely that—the collection of entities and subjects that we want to *talk* about and *investigate*, without necessarily having made a prior commitment to their existence. We only get existence out of a deduction if we have good reason to put it in as a premise. Of course, we may wish to do logic over specific sets of entities, such as perhaps the natural numbers or the items of furniture in someone's office, which are already taken to exist and to have various definite properties. But logic should preserve a general freedom to talk of the hypothetical or the fictional, and to deny the existence of entities if the facts demand that we do so.

The description of a hypothetical or fictional object may or may not include or imply self-identity. Logic does not rule out either of these possibilities. For instance, the notion of the largest prime implies self-identity by the Peano axioms, even though there is no such number. However, it is not at all clear whether the script writers who defined Godzilla intended to imply that the creature should be self-identical; that doesn't seem to have been relevant. We are entirely at liberty to define anything with any putative properties whatsover, be it Quine's round square cupola or a non-self identical movie monster. As Gaunilo realized a long time ago, no definition by itself can have any bearing on what exists. A definition can neither *make* something exist in the real world (as supposed by the confident authors of various ontological proofs) nor can it, by itself, *prevent* something from existing (even if the description of the putative entity is contradictory). In the end, as Hume indicated, the test of existence outside the lands of mathematics is always empirical.

We suggest that we can avoid the Godzilla inference, as least so far as first order logic is concerned, by taking the following two steps:

- We entirely drop the idea that (4.1) is a theorem that applies to any arbitrarily selected class of entities. Instead, any suggestion that some entity or set of entities are self-identical must be treated as an *assumption* and introduced to a natural deduction proof accordingly, unless it is specified in advance that one is quantifying over sets of entities (such as the natural numbers) which are already known to be self-identical. In such a case, a line in a deduction that says (say) 3 = 3 is justified by reference to number theory, not to =I.

- Within the object language of first order logic we treat identity in a purely syntactic manner: if $a = b$ all that this means is that a can be freely substituted for b or *vice versa* at any point in a proof.

The key new idea is that we do not take it as a theorem that $\forall x(x = x)$. Rather, we take it that identity has to be either postulated or established from other postulates for any entity or class of entities.

The purely syntactic reading of $a = b$ reflects the fact that inter-substitution is all that matters from the syntactic point of view. Most important, we can define identity this way without having to worry about deep metaphysical questions about what it means for entities to be identical. This is hardly to suggest that one should not investigate the nature of identity or the various sorts of identity that might be tenable (for indeed, our minimal syntax could be compatible with several interpretations of identity); it is simply to insist that views about the factual or metaphysical nature of identity should not be surreptitiously built into first-order logic. So self-identity cannot be taken as a given for all possible objects. Instead, a self-identity claim is to be introduced into a proof if needed by an assumption (which could be either universally quantified or of the form $a = a$ for some particular term a) or by specifying in advance that one is quantifying over a class of objects (such as natural numbers) that are already known to be self-identical.

If we want a system that works the same way as standard first-order predicate logic with identity, so that we can do all of the usual definite description problems and employ the other non-problematic applications of identity, we can take it as not a logical truth but a *global assumption* (which need not be stated explicitly in each deduction) that we reason over domains of objects that are presumed to be self-identical. The logical status of =I then becomes something like the logical status of the parallel postulate in geometry. The parallel postulate was once presumed to be either *a priori* or deducible from the other rules of Euclidean geometry, but by the early 19th century it was evident that it was a logically independent assumption about the kinds of spaces that one was dealing with. (From the Riemannian viewpoint it applies to spaces with zero intrinsic curvature.) Logics in which self-identity fails or could fail for some or all non-null entities in the domain of discourse can be called *non-Aristotelian*, by analogy with non-Euclidean geometry. A logic that differs from classical (Aristotelian) predicate logic with identity *only* in that =I is taken to be a global assumption rather than a logical truth we will call an *open classical logic*—open because it is open to the possibility that the self-identity assumption might fail for one or more members of the domain. We will show below that even open classical logics must be non-Aristotelian in one particular respect, but there could well be many possible non-Aristotelian logics, just as there are many possible non-Euclidean geometries. In this respect, logic still seeks its Riemannian synthesis.

4.4 Haecceities, or the Lack Thereof

We now want to take a different tack and examine some other motivations for considering non-Aristotelian logics, apart from a desire to avoid the Godzilla problem. We began by noting that it is not good to build metaphysics into one's logic any more than one should build facts of geography into trigonometry. On the other hand, it is also desirable to have logics that are adequate to the uses to which people frequently put language, and to the kind of natural world we seem to find ourselves in. We'll point out that non-Aristotelian logics of some sort (not necessarily classical) could well have applications in the logic

4.4. HAECCEITIES, OR THE LACK THEREOF

of fiction, in philosophy, and in physics itself.

4.4.1 In Fiction

We will not venture deeply into the logic and metaphysics of fiction, except to note that it is not at all clear that simplistic self-identity is deemed to hold for all entities treated in fiction, the movies, and literature. Walt Whitman, in his poem 'Song of Myself' (1977, p. 96), famously stated, "I contradict myself...I contain multitudes." And what of a character caught in a self-contradictory time loop in a bad science fiction story who succeeds in committing suicide by shooting his grandfather? At some points along his worldline he exists if and only if he does not exist. Only fiction, of course; but the point is simply that it is open to authors even to question the self-identity of their characters.

4.4.2 In Philosophy

Within the history of philosophy (and overlapping importantly into physics) there has been and continues to be a debate between two camps who have very different views about the metaphysics of time and change. The Parmenideans see the world as static, the Heracliteans see the world as inherently dynamic. Plato (to whose work all of Western philosophy consists of footnotes, according to Whitehead) proposed a synthesis of Parmenidean and Heraclitean views: he distinguished between the world of Becoming (the unstable physical or natural world) and the world of Being (the world of stable ideal objects grasped by the intellect). Plato stated that everything in the natural world, not only obviously changeable things such as fire and water but even more apparently permanent solid matter, was in a process of perpetual flux:

> Whenever we see anything in process of change, for example fire, we should speak of it not as *being a thing* but as *having a quality*... And in general we should never speak as if any of the things we suppose we can indicate by pointing and using the expressions 'this thing' or 'that thing' have any permanent reality: for they have no stability and elude the designation 'this' or 'that' or any other that suggests permanence. (Plato, 1971, p. 68)

Nothing in the world of Becoming is ever exactly a such-and-such and thus one can never hope to fully grasp what it is. Plato's words are open to interpretation, of course, but his view seems to suggest that objects in the natural world do not have sharp self-identity because they are always in the process of becoming something else.

Nietzsche wrote in a similar vein:

> Logic too depends on presuppositions with which nothing in the real world corresponds, for example on the presupposition that there are identical things, that the same thing is identical at different points of time... (Nietzsche, 1964, §11).

On this view, even the claim that I am now sitting in the same chair that I sat in earlier this evening is (from the logical point of view) pure stipulation. The later chair is like enough to the chair I sat on earlier for all practical purposes, so I might as well call it the same chair—but this is purely a stipulation justifiable only by its practical utility.

4.4.3 In Physics

We are certainly not suggesting that something is so just because Plato or Nietzsche said it. But such philosophical views, although imprecise and open to interpretation, are not merely quaint relics of pre-scientific or 19th century romantic thought; current professional debates on the reality of time and change turn on the same Heralitean/Parmenidean point of dispute.

The static, plenum, or block-universe view is probably the sentimental favourite of many recent physicists and philosophers of science. (Kurt Gödel was a notable block universe theorist; Yourgrau (2005).) However, because of the Indeterminacy Relations and the fundamental non-Booleanity of quantum mechanics (for an explanation of which see Bub (1997)), it is not clear that the static view is consistent with quantum mechanics (Peacock, 2006). The distinguished theorist Lee Smolin, trying to understand the basis of the conceptual roadblocks which he insists dog modern theoretical physics, remarks,

> I believe there is something basic we are all missing, some wrong assumption we are all making... I strongly suspect that the key is time. More and more, I have the feeling that quantum theory and general relativity are both deeply wrong about the nature of time... We have to find a way to *unfreeze* time—to represent time without turning it into space. (Smolin, 2006, p. 257-7)

There is still no generally agreed upon method for unfreezing time, but the questions of how to represent time, and whether or not time is real or merely a funny sort of spatial dimension, are central themes in current work in quantum gravity (the attempt to find a quantum theory of spacetime structure). If the *logic* of identity is to be of any use in talking about identity of objects in time and space, it needs to be flexible enough to accommodate our rapidly evolving picture of identity in the physical world. Conceivably, some sort of non-Aristotelian logic as we conceive of it here could be a useful tool in Smolin's project to unfreeze physics.

Paul Teller (1998) has made some very relevant observations about the way that *haecceity*—the *suchness* or *thisness* of an entity, that which presumably founds its identity—is affected by quantum mechanics. His explanation of haecceity is very helpful:

> Traditionally, philosophy has talked about an object's "haecceity" to mark the idea that an object is distinct from all others in some manner that transcends all properties in any usual sense of the word 'property.' ... let us take for granted some things that presuppose the applicability of strict identity: that names can refer "directly," that is without operating as definite descriptions; *that repeated use of the same name picks out the same referent* [our emphasis]; that repeated use of the same variable bound by the same quantifier picks out the same referent; and that sets are defined

4.4. HAECCEITIES, OR THE LACK THEREOF

> extensionally... Now, a metaphysician might ask: in virtue of what does strict identity apply to an object? Haecceities... are supposed to be some metaphysical feature, principle, characteristic, or "non-qualitative property" which answers this question. (Teller, 1998, p. 117)

As Teller goes on to explain, in quantum statistics particles do not have identities that can be tracked. To adapt Teller's example, if Bloggs has $1000 in his bank account, it does not make sense to ask, *which* monetary tokens (such as pennies) does he have $1000 worth of? All that matters is that he is good for $1000. Similarly, if there are six photons in a box, all this means is that we are good for six photons; it does not make sense to ask, *which six photons are in the box*? As Teller explains, it has been found that if one assumes that the photons have distinct, trackable identities the way pennies do, one will count them wrong (because quantum particles are permutation-invariant, unlike classical objects) and get the wrong statistical predictions. It is therefore highly questionable that quantum mechanics allows for the notion of haecceity (and thereby self-identity) in anything like the classical sense. (See French and Krause (2006) for a detailed exploration of this problem.)

There is another respect in which quantum mechanics suggests something like the Heraclitean view. Any assertion in quantum physics has to be operationally grounded; by what measurement procedure could we know that a particle is identical to itself? Well, we might have to interact with it twice, and quantum mechanics tells us that there is no clear meaning to saying that if we measure (say) an electron at a certain spot, and then a tiny fraction of a second later measure another electron at nearly the same spot, that we have detected the same electron that we detected in the first measurement. To adapt a famous phrase from Heraclitus, we do and do not observe the same electron.

At around this point in the discussion, classically-minded thinkers are sometimes moved to exclaim, "But dammit, everything just *is* identical to itself!" This can be called the *table-pounding argument* because the statement is often accompanied by pounding on a convenient mid-sized object. Unfortunately for the classical realist, quantum mechanics remains unmoved by any amount of furniture-thumping. Modern physics certainly suggests, and arguably demands, that we live in an extreme Heraclitean world of flux where self-identity cannot be asserted, or at least cannot be asserted in a classical way—except as a convenient approximation at scales where quantum effects can be ignored.

One does not necessarily employ first order predicate logic to reason about quantum mechanics, but in order for logic to be as useful as possible, in the kind of physical world we live in, it should be equipped to express facts of quantum mechanics as required and should therefore not have in-built assumptions that would conflict with quantum mechanics. In the spirit of Putnam's recommendations (1968), it is desirable to seek a way of doing classical logic that would naturally generalize to quantum logic.

To conclude this section, we quote a favourite story from Bertrand Russell:

> It is obvious that, if you think of all the things that are in the world, they cannot be divided into two classes—namely, those that exist, and those that do not. Non-existence is, in fact, a very rare property. Everybody knows the story of the two Germanic pessimistic philosophers, one of

whom exclaimed: 'How much happier were it never to have been born.' To which the other replied with a sigh: 'True! But how few are those who achieve this happy lot.' (Russell, 1956, p. 147)

We have a great deal of respect for Russell, but it is by no means obvious that existence is such a clear-cut concept in a quantum universe. Perhaps if Teller is right then non-existence is not such a difficult property to attain or at least to approximate after all, at least within the limits allowed by the Uncertainty Principle.

To summarize: self-identity holds for some idealized objects, such as the natural numbers, and it is approximately enough true not to be misleading for many mid-sized physical objects such as tables and chairs. Physics tends to suggest that it could well be simply dead wrong at scales where quantum mechanics is important, although this remains an important open question. But again, if predicate logic is to be as widely applicable as possible to reasoning about things in the natural world, not to mention fictional objects, it must not be burdened with the *presumption* that everything that can be quantified over is necessarily self-identical.

4.5 The Null Object

Now we will show how non-Aristotelian even a nearly-classical open logic has to be if it is to have enough expressive power to be useful in mathematics and daily reasoning.

Let us consider the following variation on the Categorical Godzilla proof, which we dub the *Conditional Godzilla*:

1	(1)	$\forall x(x = x)$	A
1	(2)	Godzilla = Godzilla	1 UE
1	(3)	$\exists x(x = \text{Godzilla})$	2 EI

Thus by conditional proof we have

$$\vdash \forall x(x = x) \rightarrow \exists x(x = \text{Godzilla}). \tag{4.4}$$

Here we have assumed $\forall x(x = x)$ rather than taken it as a theorem. We're on firmer ground in that respect. But we still end up with a peculiar result: assuming the self-identity of all objects in the universe of discourse apparently also allows us to prove the existence of anything in that universe, only this time not as an ersatz theorem but as a consequence of the assumption on line (1). So if we give all entities in $\mathcal{U}$ the benefit of the doubt and grant them self-identity, we are still committed to their existence—but now the dependence of the existence result upon assumption is obvious.

In part this proof is simply an illustration of the point noted earlier, that in first order predicate logic, to posit any property (including self-identity) of an entity is to imply that the entity exists. So if we give all entities in $\mathcal{U}$ the benefit of the doubt and grant them self-identity—not as a presumed logical or metaphysical truth but simply for the sake of argument—that still implies that they exist. On the other hand, non-existent things are non-self-identical, simply because no properties of any sort can *in fact* be predicated of them at all (regardless of how they were defined). But what happens

4.5. THE NULL OBJECT

if the facts of a matter demand that we deny the existence of something that we had provisionally admitted to $\mathcal{U}$?

Continue the above proof as follows:

	(4)	$\forall x(x = x) \to \exists x(x = \text{Godzilla})$	1–3 CP
5	(5)	$-\exists x(x = \text{Godzilla})$	A (An empirical given.)
5	(6)	$-\forall x(x = x)$	4,5 MT
5	(7)	$\exists x(x \neq x)$	6 Duality

By asserting on line (5) the empirical fact that a certain described entity does not exist, we seem to be forced into a bizarre existence claim anyway! One approach could be to simply not make assertions like (5) on the grounds that in classical predicate logic we hold or pretend that all names refer. But predicate logic would be greatly restricted in its usefulness if it could not assert the non-existence of a putative entity known only by a name or description. That is something that we are entirely free to do in ordinary language, as well as scientific and mathematical reasoning. As we have already noted, one of the most useful tools of reasoning is the ability to discuss something hypothetically, be it the largest prime or the gunman on the grassy knoll. What we need is a natural interpretation of the odd thing whose existence is cited in line (7). We suggest that it may be useful to think of this object, or "object", as a *null entity*. In a logical system it acts in a way analogous to the ground in an electrical circuit; it is the elephant graveyard for all names and descriptions which fail to refer.

One can see a foreshadowing of this approach in Carnap's interpretation of Frege's solution to the problem of improper definite descriptions (Carnap, 1947, 35–9). As Carnap explains, Frege was concerned to construct his ideal logical language so that a definite description picked out a unique object. There is an obstacle in the cases of improper definite descriptions, terms which have the form of a description but which name nothing or many things. Carnap observes that a possible response is "to count among the things also the *null thing*, which corresponds to the null class of space-time points" (1947, 36). It is beyond the aim of this paper to go into more detail regarding the problem of improper definite descriptions, but it is well worth noting that there is a precedent in the literature for this kind of solution to problems not too dissimilar to those in which we are most interested here.

For Carnap, a null object a_0 is simply a name that is left free—it does not denote anything. Since names can denote anything we want, it is open to us to simply leave one name unassigned in the course of a piece of reasoning. Classically (i.e., in a logic with a sharp concept of identity), a null object can be naturally defined as follows:

$$a_0 := x(x \neq x). \tag{4.5}$$

("Let a_0 be an x such that $x \neq x$.")[3] It is not essential that a_0 be defined this way: that is, as whatever is non-self-identical. The notion of a null object could easily survive

---

[3] Our notation here is non-standard and requires some explanation. It would be more common to write a definition such as this using Hilbert's ε-symbol, as $a_0 := \varepsilon x(x \neq x)$. However, the usual reading of $\varepsilon x A$ commits us to more than we want when discussing null objects: (1969, p. 1) says, "Intuitively, the ε-term $\varepsilon x A$ says 'an x such that if anything has the property A, then x has that property'." Our simpler notation is inspired by the reading of $\exists x F x$ as "There exists an F." The symbol $\exists$ is "there exists," and $x F x$ is "an F".

a loosening or broadening of the concept of self-identity. However, if we accept the classical notion of the null object, then lines (4)–(7) above do not in fact demonstrate the existence of anything at all—precisely because they demonstrate the existence of the null object, which is not anything at all. So we have not stumbled into an existence claim simply because we tried to deny the existence of something.

The move to free logic provides another natural motivation for considering the null object. Again, a free logic is defined as a system of predicate logic which allows for the possibility of empty domains of discourse and predicates that do not refer. Consider $Fa \vee -Fa$. This is true even in $\emptyset$ (the null set, an empty universe) because if a term a does not refer then $-Fa$ holds for any predicate F. (If the present King of France doesn't exist then it is true that he is not bald.) However, the fact that $Fa \vee -Fa$ is true in an empty universe seems to license a dubious inference:

(1) $Fa \vee -Fa$ Theorem
(2) $\exists x(Fx \vee -Fx)$ 1 EI (???)

That is, we seem to have once again inferred a theorem asserting the existence of an entity—this time, an apparent element of $\emptyset$!

One way to deal with this is to not talk about empty universes; and this is what is usually done in elementary predicate logic texts, where the puzzle of the empty universe is either not mentioned or glossed over. Another way to block this inference is to not allow EI for empty universes, but this requires that one know in advance that a universe is empty, and we are supposed to be able to do logic without any existential presumptions at all. Here, again, a null object can help us. Classically, we have $\emptyset = \{x | x \neq x\}$. Then $a_0 \in \emptyset$. (Indeed, it's the only element of $\emptyset$.) Then line (2) above can only apply to a_0 (in $\emptyset$), and we have (in $\emptyset$) $Fa_0 \vee -Fa_0$. Now, Fa_0 is false for any predicate F; by bivalence $-Fa_0$ and therefore $Fa_0 \vee -Fa_0$ are true in $\emptyset$. So the above deduction is valid in $\emptyset$.

Please note: this bit of reasoning does not imply that there actually is anything at all *in* $\emptyset$!

Null objects thus allow us to extend the validity of certain puzzling arguments to empty universes in a natural way, and allows us to express the non-existence of named or described entities when the facts demand that we do so—so long as we decide that we can live with one more very odd sort of mathematical creature under the floorboards of everyday reasoning. Just as set theory would be hobbled without the formal device $\emptyset$, and arithmetic could not operate without 0, it could well be that predicate logic has been hobbled all along without a formal, placeholder referent for names and descriptions that do not refer.

One further point: is the null object self-identical? Clearly not, by its very definition, Eq. (4.5). So even if we want to keep our predicate logic as close to classical as possible (by adhering to =I as a global assumption but making no other changes in our deductive methods), if we also want to be able to assert that some names or descriptions fail to refer

---

Any expression of the latter form can be called an *indefinite denotator*. Precisely because it is so thoroughly indefinite, it could (depending on the facts of the matter) denote nothing.

there must be at least one non-self-identical (and therefore null) object in every domain of reference. To this extent, then, even an open classical logic is non-Aristotelian.

We were tempted to speak of *the* null object, but Gillman Payette (private communication) was quick to point out to us that even this would be saying too much about it. Suppose that we tried to define a_0 as follows:

$$a_0 := \iota x(x \neq x). \tag{4.6}$$

("Let a_0 be *the* x such that $x \neq x$.") To speak of anything as *the* F is to allow that it can be equal to something bearing a proper name. Suppose that $a_0 = b$. So long as we are allowed =E (and we could not do much useful reasoning about identity without it), we can substitute a_0 for b and get that a_0 is self-identical—precisely the thing we don't want. So a_0 cannot even have the property of uniqueness. In this respect the analogy between null objects, and the null set and 0, breaks down because the latter entities can stand in identity relations. So the null object or objects must remain utterly indefinite; our a_0 is just a placeholder for the absence of all properties, demanded by the syntax of predicate logic.

There is one further intriguing observation to be made about null objects. The definition (4.5) is a very natural way to specify a null object in Aristotelian logic. But suppose we want to consider non-Aristotelian logics where classical identity is not always available. We would need a more general conception of null objects. If we are allowed to quantify over predicates, then we could define

$$a_0 := x(\forall F(-Fx)). \tag{4.7}$$

This has the advantage that it could apply to logics without classical identity. But one encounters a challenge that should by now be very familiar to logicians. Let us say that an object is *prediphobic* if it will admit of no predicates whatsoever:

$$Px := x \text{ is prediphobic} \tag{4.8}$$
$$:= \forall F(-Fx)$$

Then clearly $Pa \to -Pa$, and given Pa as well, we have detonation. This is a typical instance of the hazards of second-order logic. It also suggests that any logic in which classical identity fails, but in which a more general null object of the form (4.7) is desired, would have to be paraconsistent in some sense.

4.6 Summing Up

The notion that =I is a theorem or logical truth leads to the unacceptable result that the existence of any named or described entity can be proven as a matter of logic. The most natural way out of this embarrassment is to think of =I as an assumption, not a "logical truth". Indeed, there are ample reasons within fiction, philosophy, and physics why we might want to speak of entities whose self-identity is in doubt, and we should have logics that are open to this possibility. We suggest that a logic in which =I holds for all non-null entities in its domain be called *Aristotelian*; otherwise, *non-Aristotelian*.

If =I is taken to hold as a logical truth, we'll call that a classical Aristotelian logic. (We expect that such logics will sooner or later become historical curiosities.) If =I is taken to hold merely as an assumption, but still an assumption applying to all objects in the domain of discourse, we'll say that such a logic is an *open* Aristotelian logic. We conjecture—though this remains to be shown in full rigour—that an open Aristotelian logic can do all the deductive work that classical Aristotelian logic can do, without falling into the absurdity of the Categorical Godzilla. Beyond this, an important research project is to explore possible non-Aristotelian logics.

Our attempt to de-ontologize logic (by removing =I as a theorem) ironically forces us to include null objects in any domain of reference, even when the logic is Aristotelian. But this is not an expansion of our ontology (except for a modest addition to our collection of symbols) because the null object is not any *thing*, even though it may turn out to be just as indispensable as ∅ and 0.

One result seems clear: we can no more take it to be a logical truth that everything is self-identical than we can take it to be a logical truth that everything is green. If we do impute self-identity to all non-null members of a domain of discourse, it is only by courtesy or because it is a domain (such as ℕ or the furniture in someone's office) for which we have good reason to think that self-identity holds throughout. And any logic that hopes to be adequate to the ordinary demands of discourse in the real world must always allow for the possibility that self-identity fails for some entities of interest.

APPENDIX: Some Properties of an Open Classical Logic

We have a lot of work to do in order to clarify the properties of open classical logic, let alone explore other non-Aristotelian logics that might be feasible and link them with free logics. Here we list without proof some immediate consequences and properties of an open classical logic.

It is simply a result of classical first order logic that we have $a = a \to \exists x(x = a)$. For a natural deduction system, this is very easily provable by the rules of EI and conditional introduction, with $a = a$ as an assumption. If the system is stated in a normal axiomatic presentation, we have as an axiom that $B \to \exists x B$, where some instances of a particular name a occurring in B can be replaced by the variable x, bound by the existential quantifier. The identity $a = a$ is such a formula, and so $a = a \to \exists x(x = a)$ is just an instance of this axiom. A result of this with a classical theory of identity is that $\exists x(x = b)$ is provable for any name b, whereas in our open logic, $\exists x(x = b)$ follows only in cases where we explicitly assume $b = b$.

We retain the provability of sequents of the form $a = b, Pa \vdash Pb$. In the natural deduction setting, this is enforced by a rule of identity elimination. More generally, it is a form of Leibniz's law. For axiomatic and sequent calculus purposes, we can simply include this sequent as a primitive rule, as is common in the proof theory literature.

In a sequent system our proposal amounts to rejecting $\vdash a = a$ as an axiom. However, given the inclusion of the identity elimination rule, all one need to do to reason more or less as usual with identity is to include a premise of the form $a = a$; that is to include $a = a$ on the left-side of the $\vdash$. Including this extra premise is always admissible, by the

4.6. SUMMING UP

rule of thinning, and so, just as in the natural deduction case, we demand only that one make one's *extra-logical* assumptions about identity explicit.

A result of this rule is that $a = b \vdash a = a$ is provable. So, under the assumption that a is equal to anything, we have that $a = a$, and if $a \neq a$, then a is identical to nothing. With identity elimination, we can also easily prove that $a = b, b = c \vdash a = c$, where $b = c$ is the formula in which a is substituted for b to attain the conclusion. Hence, we can clearly prove that $(a = b \land b = c) \to a = c$. Similarly, we can show that $a = b \equiv b = a$. So, as a result, = enjoys symmetry and transitivity in all cases, and reflexivity in those cases where it applies at all. While this 'conditional' reflexivity is strictly weaker than reflexivity, it guarantees that in contexts where we have assumed self-identity to hold of the names we reason with, identity behaves classically. The differences are, of course, with those names of which we make no such assumption.

Of course, since we reject that $a = a$ is a theorem, we do not have that the formula $a \neq a$ allows for the proof of any formula whatsoever. In general, the assumption $a \neq a$ will only generate triviality when we explicitly assume some other formula which implies $a = a$, because we retain the rule of *explosion* (ex falso quodlibet).

We also retain the *law of excluded middle*, because the propositional fragment of our logic is purely classical. So, we have that $\vdash a = a \lor a \neq a$; however, as we do not have that $\vdash a = a$ or $\vdash a \neq a$, the logic is not *prime*. This is just to say that it is not universally true that $\vdash A \lor B$ holds only when either $\vdash A$ or $\vdash B$ holds. This leaves a potentially interesting avenue to intuitionist open logic available. It strays beyond our aims to investigate such a logic, but we note here that such an approach may have interesting consequences for common subjects which motivate intuitionistic logic, for instance mathematics[4] and other areas where epistemic restrictions on our knowledge are salient.

No proposal to amend an established logic can be taken seriously until the metatheory of the amended logic is worked out. We have yet to do this. It seems very likely that *dropping* a rule of inference leaves us on safe ground with regard to soundness. In particular, dropping =I as a theorem makes no difference at all to what can actually be deduced with first order predicate logic with identity except that we lose certain inferences that (for reasons we have explained) we would like to lose anyway.

Completeness is a more difficult question: to show that open classical first order logic with identity is complete we would have to show that any formula we can *no longer* prove (by having removed =I as a rule of deduction) is false in some models that leave true all the theorems that did not require =I. To put it another way, completeness is all about whether one can prove all of the tautologies in a system with the resources of the system. The Categorical Godzilla shows us that if the standard approach is complete and sound, $\exists x(x = a)$ must be a tautology for any term a in every possible model; and as we have noted this could make sense only if it is somehow known that every term in the language refers. In an open classical logic $\exists x(x = a)$ is most certainly not a tautology in general, but rather a statement that depends upon the facts of a case.

---

[4]Consider the fact that from ZFC alone we can prove neither $\aleph_1 = 2^{\aleph_0}$ nor $\aleph_1 \neq 2^{\aleph_0}$. Of course, giving a set theoretic analysis of = which matches our assumptions about this predicate goes beyond our purposes, but it may be a valuable way forward. This is in contrast to the well-understood notion of *bijection* underwriting = in ZFC, and other common set theories.

So again, we think that dropping =I as a rule of inference will only prevent one from being able to prove formulas that have no business being tautologies anyway. But this question requires a more thorough study.

If nothing else, this appendix makes it clear that this work is a very first step into an interesting new territory, about which almost everything is as yet unknown. Our mere hope at this stage is that the reader is intrigued by our proposal, and convinced enough by our philosophical argumentation to think that it may be worth developing in more detail.

Acknowledgements

We are grateful to the Universities of Lethbridge and Connecticut for material and financial support, and to Bryson Brown for many valuable discussions about logic and other matters. Karl Laderoute pointed us to the remarks by Nietzsche, and John Woods has provided helpful comments about the notion of identity. Gillman Payette commented insightfully on an earlier version at the Canadian Philosophical Association at Congress 2016 (at the University of Calgary). We alone are responsible for any errors or omissions that may remain in this paper despite all the good advice we've received.

4.7 Bibliography

Aristotle. *De Sophisticis Elenchis*. Online at
http://etext.lib.virginia.edu/toc/modeng/public/AriSoph.html; retrieved June, 2012.

Bell, John L., D. DeVidi, and G. Solomon, *Logical Options: An Introduction to Classical and Alternative Logics.* Broadview, 2001.

Bub, Jeffrey. *Interpreting the Quantum World.* Cambridge: Cambridge University Press, 1997.

Carnap, Rudolf *Meaning and Necessity: A Study in Semantics and Modal Logic.* Chicago: University of Chicago Press, 1947.

Frege. G. 'Begriffshift, a formula language, modeled upon that of arithmetic, for pure thought.' First publication 1879. In J. van Heijenoort (ed.), *From Frege to Gödel: A Source Book in Mathematical Logic, 1879–1931*. Cambridge, MA: Harvard University Press, 1967, pp. 1–82.

French, Steven, and Krause, Décio. *Identity in Physics: A Historical, Philosophical, and Formal Analysis.* Oxford: Clarendon Press, 2006.

Herrick, Paul. *Introduction to Logic.* New York: Oxford University Press, 2013.

Kneale, Martha and William Kneale. *The Development of Logic.* Oxford: Clarendon Press, 1962.

4.7. BIBLIOGRAPHY

Lambert, Karl. *Philosophical Applications of Free Logic*. New York & Oxford: Oxford University Press, 1991.

Leisenring, A. C. *Mathematical Logic and Hilbert's ε-Symbol*. New York: Gordon and Breach, 1969.

Lemmon, E. J. *Beginning Logic*. Indianapolis, IN: Hackett, 1978.

Nietzsche, Friedrich. *Human, All-too-Human: A Book for Free Spirits*. Part I (1878). Translated by Helen Zimmern. In *The Complete Works of Friedrich Nietzsche*, Vol. 6. New York: Russell & Russell, 1964.

Nolt, John. 'Free Logic.' In E. N. Zalta (ed.). *The Stanford Encyclopedia of Philosophy*. Spring 2011 Edition.

Peacock, Kent A. "Temporal Presentness and the Dynamics of Spacetime." In D. Dieks (ed.), *The Ontology of Spacetime*. Amsterdam: Elsevier, 2006, pp. 247–61.

Plato. *Timaeus and Critias*. Translated by Desmond Lee. Harmondsworth, UK: Penguin, 1971.

Putnam, Hilary. 'Is Logic Empirical?' In Robert S. Cohen and Marx W. Wartofsky (eds.), *Boston Studies in the Philosophy of Science, Vol. 5*. Dordrecht: D. Reidel, 1968, pp. 216–41.

Russell, Bertrand, and A. N. Whitehead. *Principia Mathematica*. Second Edition. Cambridge: Cambridge University Press, 1927.

Russell, Bertrand. *Portraits From Memory and Other Essays*. London: George Allen & Unwin, 1956.

Smolin, Lee. *The Trouble With Physics: The Rise of String Theory, the Fall of a Science, and What Comes Next*. Boston & New York: Houghton Mifflin, 2006.

Teller, Paul. 'Quantum Mechanics and Haecceities.' In Elena Castellani (ed.), *Interpreting Bodies: Classical and Quantum Objects in Modern Physics*. Princeton, NJ; Princeton University Press, 1998, pp. 114–41.

'Song of Myself,' in *The Portable Walt Whitman*. Revised and enlarged edition, selected & with notes by Mark van Doren. New York: Penguin, 1977, pp. 32–97.

Yourgrau, Palle. *A World Without Time: The Forgotten Legacy of Gödel and Einstein*. New York: Basic/Perseus, 2005.

Chapter 5

Alice Munro's "Wild Swans"

Steven Burns[1]

In 2013 Alice Munro was awarded the Nobel Prize for Literature. Her great works of art are worthy of philosophical attention for a variety of reasons. I am going to draw a connection between an Alice Munro short story and a recent book in philosophy of science. Elisabeth Lloyd's book, *The Case of the Female Orgasm*, is about explanation in evolutionary biology. It is the furthest thing from sexy. Alice Munro's story, "Wild Swans", on the other hand, is erotic and disturbing. I take it that the wild swans of her title are both a symbol of purity and a symbol of wildness. They are mentioned four times: in the title of course, and then twice with apparent innocence. There is a description of wild swans settling on a pond in the countryside, for instance. But the final image of swans bursting into flight is the image that accompanies an orgasm that a young woman experiences while travelling on a railway train. We shall have to confront this event in some detail. [I apologize to anyone who would rather not focus on an intimate subject, especially one that invokes sexual molestation. Please treat this as a WARNING.]

5.1 A Philosophical Thesis

First let me offer a philosophical thesis of my own. I have a theory that I want to put on the table right away. It is a commonplace in discussions of literature to claim that the greater a piece of writing, and the richer a work, the more interpretations it will sustain. I claim that there is a mistake in that idea. True, it is important to keep an open mind, and to try to imagine various interpretations and different ways of reading a work of literature. But some interpretations are better than others, and if that is so, then it is likely that one interpretation will be best of all. I call that the *best reading* of a work, and I think that that is what we are all looking for when we argue about how to interpret a novel or short story. In what follows I offer some steps in the direction of the best reading of "Wild Swans".

---

[1] Department of Philosophy, Dalhousie University, steven.burns@dal.ca

I don't want that to be an arrogant claim. Here is an argument that supports it. Let's accept that I am wrong. If there is no best reading, then a work can support more than one equally good reading. Consider the well-known duck/rabbit drawing:

Figure 5.1: Duck/Rabbit

It is a drawing that does sustain more than one equally good interpretation. There is no more reason to believe that the protrusions are rabbit ears than that they are a duck bill. It can be a duck, or a rabbit, but it cannot be both at once. Either explanation is as well supported as the other. I claim, however, that the ambiguity only exists because the drawing is so oversimplified, so schematic. If we were to add feathers or fur to the drawing we would make it richer, make it a better portrayal. But at the same time we would reduce the ambiguity of its meaning. So it goes with greater art. The richer and more detailed a work, such as any of Alice Munro's stories, the less likely it is to sustain ambiguity, and the more likely it is that some interpretation will prove to be the best one. That conclusion, of course, is just the opposite of the one we assumed when we began.

Why does a more detailed work reduce ambiguity? Because there is more evidence, more details of plot and character for instance, that will support one interpretation over the others. Let me try out the theory on a short example.

5.2 Introduction to "Wild Swans"

Alice Munro's story "Wild Swans" (*Who Do You Think You Are?*, 1978) has as its climax a female orgasm. A young woman named Rose is taking her first train trip by herself; she is on her way from the small town of West Hanratty to spend two days in the city of Toronto. This is a big step in her life; and so is the orgasm that she has while travelling in the train. Here is Munro's account of it:

> She was careful of her breathing. She could not believe this.... The gates and towers of the Exhibition Grounds came into view, the painted domes and pillars floated marvellously against her eyelids' rosy sky. Then flew apart in celebration. You could have had such a flock of birds, wild swans, even, wakened under one big dome together, exploding from it, taking to the sky. She bit the edge of her tongue. (122)

If this is not an orgasm, then I am profoundly mistaken about the story. But I think that it is unlikely that I am mistaken. Although it is probably not Rose's first orgasm, it is most likely her first orgasm with another person, and as such it is of considerable interest in itself. Moreover, it happens in a public place, in a railway compartment and in the presence of other passengers sitting nearby, so it is exceptional, and thus of even more interest. Finally, it apparently happens at the hands of a complete stranger and without the young woman's consent, so it is of very serious moral or ethical interest, as well. What is most striking about the story, then, is not the orgasm itself, but the disturbing issues of *who* provides the orgasm, and of how it is provided. A stranger takes the seat beside her on the train. He is an older man. Rose does not find him particularly attractive. He claims to be a United Church minister. After a while he lets his newspaper fall across both of their laps and appears to fall asleep. Rose is startled to feel his hand upon her knee, hidden beneath the newspaper. She is surprised, taken aback, but she does not react. The hand slowly moves. Rose is mesmerized, but eventually finds herself sexually aroused. Shocked with herself, she parts her legs. The hand then brings her to her orgasm. "Victim and accomplice", Munro names her (152). Neither of them acknowledges what has happened. They part with minimal courtesy, as strangers. I shall return to the story, and to some of these other dimensions of the event in the train, but first I shall discuss the orgasm itself.

5.3   Introduction to Lloyd's Evolutionary Biology

Elisabeth Lloyd has recently made a fascinating contribution to Philosophy of Biology. She titles her book *The Case of the Female Orgasm* (2005), as though she were Sherlock Holmes tracking down a mystery.[2] She defines the orgasm as "a sensory-motor reflex including clonic contractions (spasms) of the pelvic and genital muscle groups" which is experienced as a "combination of waves of a very pleasurable sensation and mounting of tensions, culminating in a fantastic sensation and release of tension" (21). (I do not need to explain how this is related to Rose's wild swans. But I do need to add an aside. Lloyd's contentment with 'very pleasurable' and 'fantastic' does prejudice some issues. There are both males and females who find at least some of their orgasms painful or disgusting, or at least embarrassing. And Schopenhauer is also not alone

---

[2] Elisabeth A. Lloyd, *The Case of the Female Orgasm: bias in the science of evolution* (Cambridge, Massachusetts: Harvard University Press, 2005). [This work should not be confused with Joan Elizabeth Lloyd's *The Perfect Orgasm: how to get it*, which is another sort of book altogether.]

in his more moderate view that sexual pleasure is trivial and disappointing.[3]) Despite these exceptional cases, I think we can assume with Lloyd that orgasms are normally sources of pleasure. Lloyd is interested in how evolutionary biology explains things, in particular how it explains features of human beings. Why human females evolved to have orgasms - when most of their primate relatives do not have them - is a mystery for evolutionary biology, she says. It serves no reproductive purpose, and so should have never appeared. And if it did appear by accident, then it should have long since been bred out of existence. The male orgasm is quite different. It is related to a build-up of quasi-hydraulic pressure which requires periodic release. And of course it plays a functional role in the reproductive act. Men who do not have orgasms during sexual intercourse will not often reproduce; the trait is *selected*. So evolutionary forces will preserve this feature of male biology, and that explains the near universality of the male capacity for more or less regular orgasm.[4]

The female orgasm is not necessary for reproduction. Yet it certainly *has* been preserved. Professor Lloyd surveys 34 studies of North American and European women which provide some empirical evidence of the frequency of female orgasm. Her "main conclusion about these studies [is that] women do not reliably have orgasm with intercourse" (42). On the other hand, almost all the women studied do have orgasms. There is considerable variation in the results of these researches, but they paint a rough picture. 90 to 95% of the women studied report having orgasms; fewer than 10% of women claim never to have had an orgasm. But those figures include orgasms achieved by masturbation and other means. The figures are quite different when the question is whether women have orgasms during heterosexual intercourse. (Here I must be explicit. By 'intercourse' she does not mean oral or anal or other forms of mutual sexual activity, but ordinary genital coitus.) Lloyd sums up the studies this way: "approximately 25% of women claim always to have orgasm during intercourse, while a narrow majority of women have orgasm with intercourse more than half the time..., [and] roughly one third of women rarely or never have orgasm with intercourse" (36).

These findings are based almost exclusively on self-reporting, which may bias the results. Further, they say nothing of the great differences among the orgasms that different women experience, and nothing about the capacity for multiple orgasms which are experienced by some 14% of the women studied (24). They tell us nothing directly about the large variation in effort required by different women, and nothing about the frequency with which various women seek orgasm; some may find it often not worth the effort, even while engaged in sex. We learn nothing about differences due to the varying competence of the males involved. Nor do these general results tell us anything about racial or cultural differences, nor about homosexual intercourse, and so on.

Lloyd makes it clear that the connection between female orgasm and reproduction is problematic. The female orgasm is not necessary for intercourse. If the orgasm is 'non-functional' in this sense, why has nature preserved it? Or, to put the puzzle another way. Some have speculated that women who regularly reach orgasm during

---

[3] E.g., at Arthur Schopenhauer, *The World as Will and Representation*, Vol. II, p. 531 ff.

[4] Both Lloyd and I are writing as though there were only the two kinds, male and female. Neither of us wishes to deny the existence of various intermediate cases, nor their importance, but that is material for a different discussion.

5.4 Crime, Coercion, and Consent

In Alice Munro's story, "Wild Swans", Rose has an orgasm. I do not want to suggest that Munro is interested in biological explanation or new research questions. However, a fair amount of what Lloyd has explained is relevant to Rose's orgasm. I think we may assume that it is likely caused by clitoral stimulation; it is not caused by intercourse. Rose's physiological responses are normal (they are more or less exactly as Lloyd defined them). And we may take it that this experience has no direct reproductive relevance. But most of what interests us about Rose's story are the psychological and moral dimensions of her orgasm. *Desire* for it is a powerful motivating force, and it dramatically influences her behaviour. She is gradually drawn past a tipping point and abandons herself to sexual pleasure. What is important for Munro is that the orgasm marks a passage - Rose is on a trip from her small town to the big city of Toronto - she is on a railway train, with its metallic heartbeat beneath her seat, the Blues rhythm of the passing of life, the travelling motif. On this trip she passes from girl to young woman. She has a new experience, an adventure; there is a new complexity in her life: a new wonder, a new pleasure, a new puzzle. All of this is consistent with the larger story of Rose.

It is helpful to know that Alice Munro published a novel (*Lives of Girls and Women*) in 1972. It consists of short stories, but they are linked by theme and characters, and it is easy to agree that the book is a novel rather than a collection of stories. The collection of stories that is known either as *Who Do You Think You Are?* or *The Beggar Maid* was at times conceived by Munro in the same way. In the end she decided that she could not sustain the novel form, because the set of stories required at least two incompatible central characters. Nonetheless, eight of the stories are about Rose.[5] Munro's New York editor, Sherry Huber, was trying to get Munro to shape the book as a novel, a collection of connected stories about the one character. The Canadian publisher "was scrupulous not to pressure" Munro. It was one result of this tension that the book appeared in the United States as *The Beggar Maid*, and in Canada as *Who Do You Think You Are?* Huber describes Rose's story:

> *The Beggar Maid* is the story of Rose, in her journey from childhood to womanhood, a trip that takes her from an ingrown, rude life in a small town, to college, marriage, motherhood, and later a separate peace as an actress.

---

[5] Besides *"Wild Swans", the other Rose stories are: "Who Do You Think You Are?", *"Royal Beatings", "Privilege", "Half a Grapefruit", *"The Beggar Maid", *"Simon's Luck" and "Spelling". Four of them (*) are included in Selected Stories. For more details of the complex publishing negotiations for *Who Do You Think You Are? / The Beggar Maid*, see the account in Robert Thacker, *Alice Munro: writing her lives* (Toronto: McClelland and Stewart, 2005), Chapter 6.

> It is an *immense* journey, a painful journey, which Rose makes alone, armed with an unwavering, penetrating sense of other people's foibles and sins. She is shy but ambitious, and gifted with a ribald, humorous sense of appreciation for the 'luck' in her life. (quoted by Thacker, 346-7)

She also has a predilection for getting herself in trouble. In the story "Royal Beatings" we meet a younger Rose. We learn that she often rejects her step-mother Flo's guidance. She is mildly disobedient, but just enough to provoke Flo. And then we learn of her tendency to let things get to a point where consequences are inevitable. Flo will call in Rose's father, who will work himself up to a righteous indignation, and then administer a thrashing. Rose can see this coming, and seems to relish its apparent inevitability, even though she is quite brutally beaten on these occasions.

If we focus on our fragment of that larger story, if we focus on "Wild Swans", we see that the trajectory of her life is a story in which this train ride to Toronto - and this particular orgasm - is just one of the passages which mark her transformation from small town girl to big city artist. It is, however, a remarkable moment in that larger story. We can make little sense of Rose's letting this moment happen unless we attend carefully to what she is coming from and where she thinks she is going. Flo is Rose's stepmother. The story begins by portraying the narrow-minded and fear-inducing warnings that Flo imposes on Rose. As Rose is preparing to leave for Toronto Flo warns her about White Slavers. An innocent looking old woman, or a man dressed as a minister, might sit beside her in the train, offer her drugged candy, and then transport her to a life of prostitution and degradation. Flo had once worked in Toronto, and had seen prostitutes, and knife fights, and worse. "That was how she knew all she knew" (116). Another character is introduced. He is a character-type who appears in other Munro stories, too. Here he is a little, bald undertaker who buys candies from Flo's store. Flo tells Rose that he uses the candies to seduce women in the back of his hearse. "All nonsense, Rose thought. Who could believe it, of a man that age" (117)? Clearly Rose does not share Flo's worries, she is eager to make the trip to Toronto on her own, and in a sense she is setting off to prove to herself that Flo's view of the world is wrong.

She is also setting off with some secret hopes. Flo does not know that Rose plans to buy some hair-remover for her arms and legs while she is in Toronto. "She also planned to buy some bangles, and a [powder-blue] angora sweater. ... She thought they could transform her, make her calm and slender and take the frizz out of her hair, dry her underarms and turn her complexion to pearl" (118). She also has fantasies about men, and "a considerable longing to be somebody's object" (118). So the event on the train can be seen as Rose's escape from Flo's vision of the dangers of the world. She does after all retain an open mind about the minister disguised as a layman (that is, he is dressed in ordinary clothes, but claims to be a pastor of a church), and she lets herself be sexually compromised, all despite Flo's explicit warnings. The event on the train can also be seen as a sort of fulfilment of Rose's hopes for the trip. Having a stranger's hand on her knee is much more adventurous than buying an angora sweater, but it is consistent with her plan to do some daring things while away from home. And we have seen that it is in her character to provoke a reaction and then 'wait for the inevitable'.

We may still be puzzled by Rose's apparent consent. She is horrified by the assault,

5.4. CRIME, COERCION, AND CONSENT

but does nothing to stop it. Why does she not say no? If at the end of the trip Rose had complained of assault, we would be inclined to side with her. We would blame the man for taking advantage of her, for exercising his authority as a minister of the church and a mature adult over a young and relatively defenseless victim. We would think that any suggestion that Rose had invited or consented to the assault was a matter of blaming the victim. Even if Rose admitted to finally wanting the orgasm, we might still insist that she did not want to be assaulted, and not by this man. So he is still an offender. We would probably want to have the man in jail. But that is not exactly the story that Munro is telling. Let us look more closely at the matter of consent.

Jean-Paul Sartre, in the famous chapter on 'Bad Faith' in *Being and Nothingness*, uses a notorious example of a man attempting to seduce a woman.[6] She is with him for the first time, is aware of his politely-disguised desire for her and enjoys his flattering attention. However, the decision she will have to face, whether or not to accept his sexual advances, is one she would rather not face, at least not just yet. When he takes her hand the urgency of the decision is pressed upon her. To take her hand away would be "to break the troubled harmony which gives the hour its charm." But leaving her hand in his is consenting to flirt; it makes the relationship explicitly sexual. With the confidence of an experienced seducer, Sartre writes:

> We all know what happens next. The young woman leaves her hand there, but she *does not notice* that she is leaving it. She does not notice because it happens by chance that she is at this moment all intellect. she draws her companion up to the most lofty regions of sentimental speculation; she speaks of life, of her life, she shows herself in her essential aspect - a personality, a consciousness. And during this time the divorce of the body from the soul is accomplished; the hand rests inert between the warm hands of her companion - neither consenting nor resisting - a thing. We shall say that this woman is in bad faith. (Sartre, *Being and Nothingness*, pp. 55-56)

Sartre goes on to analyse the techniques by which bad faith can be achieved: he proposes for instance that key terms are 'metastable'. They have more than one meaning, and the meaning may appear to shift. That is, the young woman may think 'I am not leaving my hand in his', which is true because she is not actively and deliberately performing that action. But then she will slide from that meaning to the apparently implied thought, 'he is not holding my hand in his', which is just false. On one analysis, maintaining such a metastable condition entails holding incompatible beliefs simultaneously, but not acknowledging their incompatibility. On this view she would not explicitly 'think' either of the metastable thoughts.[7] The question of how to analyse

---

[6] Jean-Paul Sartre, *Being and Nothingness: an essay on phenomenological ontology*, tr. Hazel E. Barnes (New York: Citadel Press, 1956), Chapter II, "Bad Faith".

[7] On another analysis it involves "shift[ing] continually from one pole of the contradictory standpoint to the other." This phrase is taken from W. H. Walsh, *Hegelian Ethics* (London: Macmillan, 1969), p. 31. Walsh is discussing Hegel's criticism of the Kantian moral standpoint: "It has to be allowed that morality is committed both to the destruction of the natural passions and to their preservation" (*op cit.*, p. 30) because it needs them as temptations to struggle against. Hegel's term for how this dissimulation is accomplished is

such apparently contradictory states of mind is fascinating. Is it the paradox known as self-deception? I do not think so. We can understand Rose, I think, without having to resolve the paradox of how she can both know and not know what she is doing. I am content to say that she goes through a dawning of consciousness. Self-deception involves not-knowing; this is not Rose's case. She begins by being reluctant, then she is skeptical, but in the end she knows what is happening.

Look more closely at the sequence of events. Munro has given us seven distinct stages of awareness. (1) The man's newspaper brushes her leg. She notices this but ignores it. (2) Later the man's hand touches her knee. She does not know whether it is the newspaper or his hand. She thinks that it cannot be his hand - that would be unheard of - but she is reminded of men's hands. She has looked at them before, and wondered what it would be like to be handled by them. (3) Then it becomes clear to her that it is the man's hand. She could have "set his hand firmly on his own knee. This solution, so obvious and foolproof, did not occur to her. And she would have to wonder, Why not" (121)? Presumably she would wonder about that much later. At the moment, she finds the hand unwelcome, it makes her feel slightly disgusted. (4) But the hand begins moving very slowly on her thigh. After a time she feels "a faint wandering nausea," that turns to the humiliating recognition that she is sexually aroused. "[H]is stubborn patient hand was able, after all, to get the ferns to rustle and the streams to flow, to waken a sly luxuriance." (5) Still Rose would rather this were not happening. She wishes he would take his hand away. Her legs are still crossed. "While her legs stayed crossed she could lay claim to innocence.... Her legs were never going to open." [Here we notice the passive voice. Munro does not write, "While she kept her legs crossed ." Like Sartre's young woman, she is not taking responsibility for what her legs are doing.] (6) And what happens next? Still in the passive voice: "Her legs were never going to open. But they were. They were" (121). Now Munro shows how Rose's awareness grows. The parting of the woman's thighs is the decisive, the transcendent moment in the common sex act. Rose acknowledges that she is giving the hand access. Opening her thighs, "she would make this slow, and silent, and definite declaration" (122), writes Munro. (7) Now the man's fingers "would go powerfully and discreetly to work." Rose would "ride the cold wave of... greedy assent.... She was careful of her breathing" (122). Munro has clearly delineated these seven stages of increasing awareness and acceptance of the approaching arousal and orgasm. We might say that seduction is a closer paradigm than rape. But this is dangerous territory. Many rapists have persisted against an other's protests in hope that their victim would give in to pleasure and thus consent. But "No" does not mean "Maybe"! I do not think, myself, that Rose has consented to be molested. She has consented to the desire for orgasm, which is quite different.

Elisabeth Lloyd did not offer us detailed information about this type of orgasm, neither by intercourse nor by self-masturbation, but by what was at one time called 'petting' - masturbation by another person. Its varieties are enormous, and it would not have served her purpose particularly well. She only needed to distinguish heterosexual intercourse from all other forms of sexual activity. But if we make the distinction, we

---

"*Verstellung*". See *Phenomenology of Spirit*, tr. A. V. Miller (Oxford: Oxford University Press, 1977), pp. 374-83.

can consider a drastically different reading of the Munro story. When I presented this topic at the University of Vienna, one of my most attentive students raised his hand. "Back at stage 2, he said, "when I read: 'She thought for some time that it was the newspaper. Then she said to herself, what if it is a hand?' (149), I knew that the rest of the episode was in her imagination."[8] I was not ready for that. I had briefly considered it on first reading, since the alternative was so shocking. Perhaps the man beside Rose really was sleeping. She imagined that the touch on her knee was the touch of one of those men's hands about which she had fantasized. The rest is her erotic imagination. But surely that could not be right, I thought. What about her orgasm? Did she imagine that? This seems to me implausible, but it is easy to think that after imagining the man's hand on her thigh for a while, Rose became aroused, let her own hand slip beneath her coat, and allowed her own hand to do the disembodied work. On this reading there was no molestation, no violation, just a modest self-indulgence. Of course this is a plausible alternative reading! What was I to say?

In the event, I assumed that one of us must be right and the other wrong. I cast about for evidence to support my own reading. Rose could have "set his hand firmly on his own knee". "His stubborn patient hand", writes Munro, "...was not, or not yet, at all welcome", she continues. Surely these phrases confirm that the man really was involved? Do they not *prove* that that interpretation of the story better explains the evidence of the text? Well, perhaps we must admit that Rose's erotic imagination might be powerful enough to override all that evidence. Moreover, those phrases are part of a *literal* reading of the story. If we are to take the story so literally, why has Munro included so much detail about the causes and contents of Rose's imagination? There are Flo's warnings, the co-incidence of the character claiming to be a minister, Rose's imaginings about men's hands. Perhaps at a *non*-literal level we should understand that the story is obviously about Rose's private fantasies, and not about the reader's. But we must also ask, would a story about Rose's imaginings be as captivating a story as one about her not rejecting the advances of a stranger?

5.5 Conclusion of Lloyd's Project

Lloyd goes on to criticize various attempts by evolutionary biologists to answer those questions at which we paused: questions about why the female orgasm has survived at all. The main contenders are adaptationist views. 'Adaptation' is the view that this trait or feature of human females constitutes a sort of 'fitness'; the orgasm, that is, makes it more likely that women will reproduce, it gives them some advantage over women without this feature or capacity. She discusses in detail eighteen versions of this hypothesis, and explains why there is no evidence to support it. A second sort of explanation is called 'non-adaptationist'. In these cases the view is that women have orgasms because it is so important that *men* do. The physical structures that are vital for male ejaculation are partly formed in the embryo before the differentiation of male and female takes place. So women have 'vestigial' capacity even though they do not need it. [A similar explanation can be given for the persistence of nipples in men. The

---

[8] I thank Mr. (now Dr.) Stephan Köstenbauer for making me rethink this point.

basic structures are formed before sex differentiation in embryo. The survival of those structures is selected because nipples are needed by women if their infants are to survive. So men have them as a useless by-product.]

I shall present two cases to help show how Lloyd reasons about these biological explanations. First is a well-developed adaptationist account by Desmond Morris.[9] He maintains that the female orgasm helps to assure that a mating pair, a man and a woman, will stay together, form a "pair bond". This in turn will tend to insure that the woman and child will be cared for by the man during the long time it takes for a human child to reach maturity. It will also tend to ensure that the woman will be faithful while the man is away from her, on a hunting trip for instance, thus strengthening the man's desire to see the child mature. The importance of pair-bonding when children take so long to mature helps to explain why the female orgasm is a human characteristic not shared with our nearest primate relatives, Morris claims.

Lloyd spends a dozen pages (49-60) detailing the mistakes in this familiar kind of explanation. One mistake is that some other primate females do experience orgasms. Stumptail macaques [a kind of monkey], for instance, have been shown to have physiological symptoms and a "complete orgasmic behavioral pattern" (54), although this happens most often when a female mounts another female. Second, Morris "seems to generalize from the male response to the female" (51) while ignoring differences. So he claims that orgasm is followed by a period of exhaustion and relaxation, although there is considerable evidence that women, unlike men, often feel "strong, and wide awake, energetic and alive" (52) after orgasm with coitus. Third, Morris thinks that orgasm helps to explain why human females will stay with one mate. A macaque female in estrus (in heat) "may copulate with 20 males a day" (54). For the human female, bonding with a single male would decrease the chances of pregnancy, so the female orgasm leads humans to copulate more often with the single partner, thus counteracting the radical decrease in the number of sperm donors. And it should increase the chance of the male's helping with the lengthy child-raising - which is not an issue for the macaque. Lloyd, however, thinks that the reasoning in this step is obscure. She is not convinced that having orgasms with intercourse would make the macaque less likely to get pregnant (nor for that matter to be less promiscuous). Fourth, Morris assumes that "female orgasm is tied to intercourse" (56). He does not even consider the point that Lloyd has carefully established, that in human females orgasm is not reliably associated with sexual intercourse. "This fact is particularly damaging; suppose females got attached to having orgasms, were not satisfied by the usual brief copulation with one male, and sought out other sexual contacts to satisfy themselves - this would work against the pair bond, and on Morris's scenario, would be selected against."[10] Why should the female orgasm not work *against* the pair-bond rather than *for* it? For

---

[9] Desmond Morris, *The Naked Ape: a zoologist's study of the human animal*, (New York: McGraw-Hill, 1967).

[10] P. 57. Lloyd then discusses another of Morris's arguments, that female orgasm encourages a prone position, and the following exhaustion gives time for the flow of sperm downward toward the womb, thus increasing the chance of pregnancy. But this is contradicted by two facts: (a) a woman can more easily reach orgasm during intercourse when she controls the friction from a superior position - the woman on top; and (b) a woman who is "strong, and wide awake, energetic and alive" after orgasm is less likely to lie still to let gravity do its work on the semen. (*Ibid.*)

5.5. CONCLUSION OF LLOYD'S PROJECT

various reasons, then, Lloyd finds the adaptationist accounts of the female orgasm to be inadequate and implausible. Often this is because the unacknowledged and biased *assumption* is that the female orgasm is essentially like the male one, and must serve a similar reproductive purpose.

In her fifth chapter, Lloyd discusses a particular *non*-adaptive account - the account given by Donald Symons.[11] His claim is that female orgasm is not an adaptation, and is not subject to direct selection pressures, but is a potentiality that is variously actualized in different cultural and individual circumstances. The existence of this potential is biologically due to the way the embryo develops, and it is a by-product of the evolutionary selection pressure on *male* sexuality. In the early stages of development, both male and female embryos have the same physical characteristics. After about 8 weeks of gestation, a release of the hormone testosterone will cause the embryo to develop into a male. Without this added hormone, the embryo will develop into a female. It is important that the male's penis and the female's clitoris are the same part of the human anatomy, they are 'homologous' organs. "Similarly, the nervous and erectile tissues involved in orgasm in both sexes arose from a common embryological source" (108). Thus the male and female orgasms "are remarkably similar" (109). But the female one does not contribute to reproductive success. It persists as a by-product of the male orgasm, which does so contribute. Whether this capacity is activated or not in a particular woman is a function of circumstances, not of evolutionary selection. In the extreme case, societies which practice clitoridectomy (so-called 'female circumcision') deliberately eliminate the capacity itself.[12]

This explanation has several advantages. (1) It explains why there are such different patterns of sexual response in women when intercourse is compared to masturbation. Orgasm during intercourse is not selected for and is not, as it were, biologically enforced. (2) The high degree of variability in rates of orgasm supports the claim that "no directional selection pressure has effectively shifted the population towards female orgasm" (136). Normally selection pressure would be expected to shift the characteristics of a population to a fixation point. (3) The cross-cultural evidence suggests that whether the potential for orgasm is activated is not dependent on biology, but on cultural influences. E.g., in cultures where men do not know of or believe in the possibility of female orgasm, and where sexual behaviour is narrowly focused on the male orgasm, women do not have orgasms with intercourse as frequently as they do in contrasting cultures. This point is supported mainly by anecdotal evidence, and it is one of the areas in which new research is particularly needed, claims Lloyd.[13] Finally (4) it explains the data from non-human primates, some of whom seem to have a potential that is largely unactivated in nature. As a result, Lloyd favours this explanation above the many adaptive ones that she criticized.

Among the objections to this sort of account, the most interesting are those of

---

[11] Donald Symons, *The Evolution of Human Sexuality* (New York: Oxford University Press, 1979).

[12] 'Deliberately' may seem an inappropriate term for such a traditional practice, whose purpose is more baffling than intelligible. If we consider orgasm essential male, however, and thus consider women who climax during intercourse as oddly masculine, then clitoridectomy makes women equally 'female'. The practice might have such an intent.

[13] Lloyd, *op. cit.*, p. 257.

feminist critics. One might expect this, for feminist philosophy has had a transformative effect on contemporary philosophy in English (and some other European languages). The first thing to note is that Lloyd's own work belongs in the strong tradition of feminist work on philosophy of science.[14] Her subtitle, "Bias in the Science of Evolution", is specifically aimed at gender bias. Androcentric (male-centred) assumptions bear considerable blame for the misdirected research and implausible reasoning that she exposes in the biological literature. But that of course does not shield her from further criticism. Feminist critics have complained of Symons' bias toward reproductive sex; they claim that he leaves out other possible selection pressures during the course of human evolution. Lloyd thinks that this is just wrong: Symons *does* emphasize the separation of orgasm from reproductive sex. The same feminist critics have complained that treating female sexuality as a male by-product is denigrating or humiliating to women. To this, Lloyd replies that "there is no reason at all to think that only directly selected traits are 'important'" (142). As an example she cites refined musical ability, which is highly important, but clearly not selected for by evolutionary pressures (or we would all have more of it).

Elisabeth Lloyd's book, then, is one discussion, clearly important to philosophy, that emerges from taking the female orgasm seriously. She has shown the inadequacy of many biologist's explanations of female orgasm, she has shown how some of those errors stem from androcentric biases, and she has shown how another account is more consistent with the data and suggests new research questions. I intend to return to questions about finding the best explanation for the evidence.

You should now notice that this paper, which seems to be about two very different texts, has taken a turn toward unity. Remember my philosophical thesis: that the best reading of the story will give the best (most coherent, simple, unique, and appropriate) explanation of the huge variety of evidence that the story provides. Just when it seems that there might be an insoluble question about whether Munro's story is best read one way or another, we should be reminded of Elisabeth Lloyd's work. She was engaged in a serious struggle about whether one hypothesis or another provided the best explanation of certain empirical phenomena. She carefully considered adaptationist and non-adaptationist accounts of the available information. She concluded that a non-adaptationist account gave a more satisfactory explanation of the data that she had to deal with. I am engaged in a parallel investigation with regard to Munro's story. I am trying to fit various proposed interpretations of the story to the evidence which the story provides. I have for the moment reached an impasse. Two readings seem equally supported by the available facts. I am reminded of the Duck-Rabbit. I shall now try to escape this unsatisfactory impasse.

5.6 Conclusions About the Story

Munro does not tell us the answer. And there is much more to be said about competing readings and ambiguous texts. Perhaps we should think of "Wild Swans" as an example

---

[14] I am thinking of work by Evelyn Fox Keller, Helen Longino, Kathleen Okruhlik and Alison Wylie, to name just a few.

5.6. CONCLUSIONS ABOUT THE STORY

of a story in which the author deliberately leaves such a question open. Perhaps she has tried carefully to put as much evidence for the one reading as for the other into the text. If she has succeeded in this, then we might be tempted to say that the best reading of the story is the one that highlights the ambiguity. And then, as another of my Viennese students insisted, it does not matter which reading is adopted; the text does not favour one over the other. [15] This seems to me to be a risky position, too, however. It leaves some of our best questions entirely unanswered: Does Rose think that she has been molested? that the man in the train is a sexual predator? or does she just indulge what she knows is a fantasy? is the man innocent?[16] For purposes of this talk, I am going to return to my original reading of the story, and assume that the stranger on the train was not innocently asleep. I have given considerable textual evidence that invites us to read the story this way, and pending a more sophisticated discussion of the possibility of ambiguous texts, I think I am justified in concluding that this is a story of a stranger giving Rose an orgasm. But I want then to go on to claim that this is not Munro's main concern.

If we think that the central moral question here is 'Did she give the man her consent?', we have to be satisfied with a mixed answer. At the beginning there was no consent at all. Had he asked her, of course she would have said no. But she also gave no sign of resistance. She could not successfully claim in a court of law that he had acted *against* her will. But he could be charged with acting without her active consent. Perhaps he did act without her consent, but she gave no sign to him that she wished he would stop. Her will remained unexpressed. However, by the time she makes her "definite declaration" by parting her legs there is a definitive answer. She has given her consent at least to the orgasm, though not perhaps to him. Anyone who has read feminist literature over the past half-century now must know that there are many cases of rape, of what should properly be called rape, which look very similar to this case; at least they look similar to the person in the man's role. The male can think that his seduction has been successful, when what he has achieved is the rape of a partner who has not given consent. On my reading, our 'sleeping' Minister has committed a crime. Rose could have him charged. But I think that Munro is not telling either one of those stories. Rose is a different case again.

Nerve

There are many other subtleties that a reader will notice in "Wild Swans". The swans themselves, for instance, are introduced first in connection with the undertaker whom Flo thinks is a womanizer. Flo imitates his contented song: "Her brow is like the snowdrift / Her throat is like the swan..." (145). Later the undertaker's surrogate, the minister in the train, tells Rose that he had "an unusual experience the other day.... I saw some [swans] down on a pond.... A whole great flock of swans. What a lovely

---

[15] I am grateful to Ms Isabella Gradinger for pressing this point.

[16] Let me note in passing, for discussion on another occasion, that the point at which our readerly minds seem to find numberless unanswered and unanswerable questions is not a random point. It is deeply embedded in the story. Rose is not in a canoe on a fishing trip! And within the story's tight constraints we may be able to find in the text good reasons for limiting even some of that apparently open-ended questioning.

sight they were... I never was lucky enough to see them before" (148). This is of course both a bit of 'innocent' conversation, but it is also a premonition of the orgasm; and it must be the source of Rose's climactic image: "You could have had such a flock of birds, wild swans, even, wakened under one big dome together, exploding from it, taking to the sky" (152). Another subtlety concerns the psychology of the man. Consider this sentence: "[S]he would make this slow, and silent, and definite declaration, perhaps disappointing as much as satisfying the hand's owner." The completion of that sentence gives sudden reality to the man. Rose has been dealing with a disembodied hand, but the man has been touching her with intent. And now we ask whether he simply intends *her* pleasure; we ask what kind of pleasure *he* derives from this touching, this molesting of a young stranger; we ask whether part of him is testing her decency and is disappointed or disapproving when she does not recoil and stop him. That one short phrase places his self-consciousness squarely within her self-consciousness. It is no wonder then that we are later told that although Rose never saw the man again, "he remained on call, so to speak, for years and years, ready to slip into place at a critical moment, without even any regard, later on, for husband or lovers" (152).

This detailed survey of what Munro tells us about Rose's growing awareness of the event and its implications does not, I said, require us to solve the problem of bad faith or self-deception, namely that she must know and not know at the same time. Rose gradually grows more aware of the unusual facts, and gradually takes responsibility for letting things happen, and then acknowledges her desire and her "greedy assent". This is why I have emphasized the passage, the growth, the transformation that this story presents. And I think that it also shows that we do not need to answer definitively the question: was she molested? or did she pleasure herself? The transcending question is that she embraced an unexpected and 'dangerous' adventure. I want to conclude by suggesting that even if we were to think the story ambiguous about whether Rose was assaulted or just fondled herself, Alice Munro does not seem inclined to answer our question, because that is not what most concerns her. The story is a story of Rose's growth; she has accepted the life-altering orgasm in the railway carriage, regardless of whether the clergyman forced it upon her or not.

The most important of all the clues to the interpretation of this story as one of Rose's growth is the way she conceives of her final arrival in Toronto. She is, as it were, still getting messages from Flo. The train station reminds her of Flo's story of a Toronto girl, Mavis, who worked in a coffee shop. Mavis had looked like a certain movie star. Once Mavis had gone on holiday. She bought herself Hollywood-style clothes and registered at the hotel using the movie star's name. "She could have been arrested, Flo said. For the *nerve*" (153). [Some of Munro's editors and advisors preferred the title "Nerve" for this story. It is easy to see why; it is a key to Rose's character and helps to explain the extraordinary event that happens to her. Munro's title, however, emphasizes the unity of the story and the quality of the experience. I think she made the right choice of title.]

So Mavis had nerve. Now the character in the main story who dressed in a disguise was the minister. And it was he who had the nerve to put his hand on Rose's knee, and see how far he could go. But that is not the most important analogy here. *It is Rose who had the nerve.* The nerve to plunge into an unexpected adventure. The nerve to ignore her step-mother's warnings. The nerve to be transformed into a different person.

The words she uses about Mavis apply directly to herself. They are the last words of the story: "She thought it would be an especially fine thing to manage a transformation like that. To dare it; to get away with it, to enter on preposterous adventures in your own, but newly named, skin" (153). That, I think, is exactly what she has done.

And that is my attempt at a best explanation of a very subtle work of art.

5.7 Bibliography

Burns, S. (1987). Sex and politics. In J. Boulad-Ayoub (Ed.), *Politique et culture, idéologie et verité*, pp. 11–15. Montreal: Presse de l'UQAM.

Burns, S. (2014). Politics in beautiful losers. In J. Holt (Ed.), *Leonard Cohen and Philosophy: various positions*, pp. 139–153. Chicago: Open Court.

Lloyd, E. A. (2005). *The Case of the Female Orgasm: bias in the science of evolution*. Cambridge, Massachusetts: Harvard University Press.

Morris, D. (1967). *The Naked Ape: a zoologist's study of the human animal*. New York: McGraw-Hill.

Munro, A. (1998). *Selected Stories*. Toronto, Canada: Peguin Canada. pp. 143–53.

Sartre, J. (1956). *Being and Nothingness*. Distributed by Random House.

Symons, D. (1979). *The Evolution of Human Sexuality*. New York: Oxford University Press.

Thacker, R. (2005). *Alice Munro: writing her lives*. Toronto: McClelland and Stewart.

Tuana, N. (2004). Coming to understand: Orgasm and the epistemology of ignorance. *Hypatia 19*(1), 194–232.

Chapter 6

Socrates the Janus-Faced:

The *Logos* of the Phaedrus

Leon McQuaid[1]

The Phaedrus is an odd dialogue. It has a peculiar setting, Socrates behaving badly, three somewhat weird speeches, and many hotly debated theories as to what exactly the dialogue is about. Thankfully we don't need to know the whole of it in order to examine its parts. For the purposes of this paper I will focus only on the logic inherent in its three speeches. 'Logic', however, is a loaded word. By 'logic' I don't mean anything formal. Instead, I only wish to analyze the dialogue in terms of various informal definitions of logic ('*logos*' to be exact). I do this to compare ancient and modern notions of logic, and to give the reader a better understanding of what I believe Plato is trying to accomplish with these three speeches. '*Logos*' has many senses and I believe that Plato intended to weave these meanings into the Phaedrus. I will untangle these meanings while also showing how entwined they are with ethics (i.e., *ethos*) and psychology. Lastly, I will suggest that ethics and logic may be inseparable for Plato and offer for discussion if this is still not the case.

6.1 The Logic of '*Logos*'

What is '*logos*'? I must confess that I don't know. Thankfully I am in good company. It is a heavy word, and one which not only has many meanings but has gained and shed various colorings throughout the centuries. Medieval scholars, for example, have gravitated towards such Biblical passages as John 1:1 'in the beginning there was the Word, and the Word was with God and the Word was God.' 'Word' in the previous passage is translated from the Greek '*logos*'. Such scholars saw Jesus as the Word of God. '*Logos*', also having the connotation of 'reason', led to lively discussions as to whether or not God was insane or irrational when Christ walked the earth. After all, if Jesus is God's reason, and Jesus was separated from the Father, would He then, be '*alogos*',

---
[1]Private Scholar, leonpmcquaid@hotmail.com

or without reason, until his Son's return? 'Reason' and 'word' were connotations of *logos* which were also at home with older Greek usages. If pressed for a provisional definition of *logos* I believe 'reason' would be the best application; but, it is more of a Procrustes' bed then a Cinderella's slipper. Yes, *Logos* meant 'reason' and 'word', but it also meant 'to collect', 'to tell', 'to recount' or 'give an account' (Braun 10). Things are murky in the past. Perhaps, then, we'd be better off sticking with more modern notions of logic.

Sadly, I have only a slightly better understanding of the contemporary incarnation of *logos*, in formal or analytical logic. Modern logicians refer to this study as the science of inference. This is a good place to start. By 'the science of inference' they mean that the study of logic is the study of deriving, quite mathematically, conclusions from premises. It is the formalization of natural language into a structure which displays the movement of reason in a predictable, mechanical way. Like mathematics, this study is vastly more sophisticated than it was in ancient times; and, like mathematics, the analytic study of logic is still fundamentally unchanged. Aristotle is credited with the invention of formal logic and you can see a scientific quality in his famous syllogism of Socrates' mortality. He wrote that we can derive the fact that Socrates is mortal, given the premises that 1) all men are mortal and that 2) Socrates is a man. Put another way: if Socrates is a man and if all men are mortal, then Socrates must be mortal. Voilà, behold the beauty of logical necessity.

One might say that Plato's logic is comparatively unrefined but this would be entirely inaccurate. Quite the contrary, Plato's dialogues represent an apex of artistic refinement. Yet, as art, it must be said that they are neither exacting nor formal. Still, Aristotle and Plato are not necessarily at odds. It is difficult to compare the two in that our knowledge of Aristotle comes from reading his lecture notes while Plato presents his thought through dramatic dialogue. Reportedly, Aristotle also crafted fine dialogues. Unfortunately, none have survived. Likewise, Plato owned an academy and may well have written notes which are lost to us. Thus, contrasting these two is precarious in light of how little we know. Personally, I believe that both preferred dialogue and were wary of how easily the written word can be distorted. I also doubt that Plato would have any ideological issues with Aristotle's formulations. I count Aristotle's insights on logic as in keeping with the general Greek appreciation for seeing a kind of geometrical harmony and beauty inherent in the natural world. Likewise, the derivation of a conclusion from its premises is the same as deriving the missing length of a right triangle.

So why do I begin with Aristotle rather than Plato? Primarily to link Aristotle to our modern science of inference and thus create some clarity in contrasting Plato to ourselves. We consider Aristotle as the father of logic and often talk of the previously stated syllogism as a kind of groundbreaking insight. Perhaps this is true, but it does not imply that Aristotle conceived of logic in fundamentally different terms than Plato. Plato's logic is squarely about verbal reasoning and less about formalization. As for Aristotle, I take his formal rules of reasoning to be more of a special use case than evidence of stark difference. Still, it's useful to use him as a foil; if only for pedagogical shorthand, to establish an opposition between what I'll refer to as formal logic and informal logic, or analytical logic and *logos*.

6.2 *Logos*: Inference, Memory and Discourse

As for inference, it is at least spiritually present in Plato's dialogues. We can see this especially in the *Meno*. There, Socrates demonstrates that a slave boy seems to have an innate knowledge of geometry. From the onset, the slave claims to have no knowledge yet Socrates leads him through simple claims to complex answers. Socrates never makes any statements about what geometry is or how its rules work, yet the boy arrives at the correct answers. We moderns may conclude that he has shown a clear ability to reason and infer. Yet, Socrates doesn't make these exact claims. Instead he insists that this ability to reason is evidence that the boy already understood geometry and only needed to refresh his memory. This framing of the event places intuition at centre stage. The slave understood the topic tacitly or intuitively prior to Socrates' prompting. Logic, or reasoning, only drew out this pre-existing understanding and made it explicit. Logic, then, is merely a tool for reminding; a memory aid.[2] This process looks remarkably the same as inference. In fact, though Socrates' epistemic assumptions are fundamentally different than our own, his use of logic is practically the same.

These epistemic differences may seem trivial but they greatly matter. The *Meno* casts Socrates as making some very uncharacteristically strong convictions on the nature of knowledge and opinion. So strong that he says:

> If I claim to know anything else—and I would make that claim about few things—I would put this down as one of the things I know. (Meno 98b)

This quote comes after explaining the difference between knowledge and true opinion (Meno 97a-98b). There, Socrates explains that there is no practical difference between knowledge and true opinion. By knowledge, he means something like being acquainted with an object—to know something is to have directly encountered and experienced it. True opinion, is more like rote understanding or knowing something without having any experience in the matter. Though these two kinds of knowing are very different Socrates argues that, in a utilitarian sense, they are exactly the same. His case is illustrated through metaphor. He states that if one wished to know the way to Larissa it would make no difference if one asked someone who had been there—whose directions are backed by experience—or someone who could simply parrot the correct directions. But this is not to say that knowledge is of the same absolute value as true opinion. As Socrates puts it:

> For true opinions, as long as they remain, are a fine thing and all they do is good, but they are not willing to remain long, and they escape from a man's mind, so that they are not worth much until one ties them down by [giving] an account of the reason why. (Meno 98a)

Here Socrates conceptualizes reason as like a map which organizes, or partitions real objects and experiences. Socrates, always cheeky, likens reason to that thing which

---

[2] In *the Phaedrus*, Socrates refers to writing as a 'tool for reminding'. He clearly sees reason as a powerful maieutic tool. Thus is writing also a maieutic tool? I think so, but I'll forego analysis for the sake of brevity.

organizes wayward ideas and ties them down like naughty runaway-slaves. Logic is fundamentally a mnemonic device. But, perhaps more importantly, the difference between knowledge and true opinion is the addition of logic. For, without logic (or reason), the facts which it maps are confused, disjointed and perhaps even mute in their nascent, intuitive form. Socrates insists that a prerequisite to knowledge is that it is teachable. It would be difficult or impossible to teach someone who doesn't have an explicit knowledge of what they are talking about. Thus knowledge must be logical. Not in the sense that knowledge can be derived out of thin air, but in the sense that logic is a prerequisite for communicability. Thus Socrates implies social, linguistic and psychological dimensions to knowledge as well as that which makes knowledge possible—*logos*.

This focus on communicability fits well with translating '*logos*' as 'speech', 'word' or 'reason'. Eva Braun in her *The Logos of Hericlitus* confirms this aspect of how the ancients conceptualized '*logos*'.

> A first survey of the activity that *logos* spans yields: collecting and laying-down, tale-telling and relating, counting and account-giving, arguing, and, so, speaking, saying, and above all, the thinking, the reasoning, that is behind uttering [...] (Braun, 10)

'*Logos*', during Hericlitus' time (535-475 BC) was fundamentally linguistic. The Phaedrus is thought to have been written roughly one hundred years later, in 370 BC. Semantics can drift significantly in one hundred years but I'm unaware of any drastic changes in the meaning of the word and I find this linguistic approach too fitting of Plato's approach to dismiss. Logic in Hericlitus', as well as Plato's time was thought of far more discursively than our own. *Logos* is fundamentally narrative or linguistic in nature.

Yet, before I get into how Plato's concept of *logos* is more language oriented than our own, I find it fruitful to speculate on some of the more archaic meanings of the term. The oldest definition of '*logos*' is 'to gather'. Braun speaks of this oldest known definition thusly:

> [...] to pick up and lay down or lay by, that is, to collect, hence to count up, to tell (as does a bank teller), to re-count (as in a tale), and thus to give an account. (Braun, 10)

Though a strange phrase, 'to collect', may only be superficially foreign. Braun does not assign a date to when this definition was used and I have no source to confirm that Plato was aware of it. However it does sound uncannily Platonic. 'To collect' connotes method or procedure: one cannot gather anything without knowing what kind, class or category of things something belongs to. This all sounds full of objectivity, rule-following-ness and good logic-y bits; and I consider it no coincidence that talk of 'picking up' and 'laying down' is so reminiscent of Socrates' method of Collection and Division. Socrates and Plato may have had little notion of an official science of inference, but they did have notions of kinds and relations. Indeed the first analytical function

6.2. LOGOS: INFERENCE, MEMORY AND DISCOURSE

of the word '*logos*' was developed as early as Pythagoras' time. The Pythagoreans discovered the number relations between octaves and harmonies. They spoke of '*logoi*' (the plural of *logos*) when determining the mathematical relationships between natural phenomena and the ratios between quantities (Braun, 30-32).

Plato's use of logic is generally more linguistic than our own. It's not a cut-and-dried case of following definite rules in order to arrive at assured conclusions. Yet it does retain a certain process of discovery which binds facts within the chains of reason. The more salient feature of this more language-oriented conception of *logos* is its reliance on speech. Language is a social phenomenon, requiring at least two speakers—requiring dialogue. Thus *logos*, in demanding expression, butts up against social and psychological concerns. To give an account is to bring to mind and thus to indirectly honor the topic at hand. As Crowley's and Hawhee's *Ancient Rhetoric for Contemporary Students*:

> In archaic Greek thought, a person's *logos* was her name, her history, and everything that could be said about her. Another word for *logos* was *kleos*, 'fame' or 'call.' Thus to be '*en logoi*' was to be taken into account, to have accounts told about one, to be on the community's roster of persons who could be spoken, sung, or written about. (Crowley and Hawhee, 12)

Logic as 'an account' may also seem odd at first blush and it's safe to say that if an English speaker has ever conflated the concepts of 'renown' and 'reason' it was a novel event. But the recognition of deeds is not a particularly Greek phenomenon. Though we now live in a 'shock culture' it wasn't long ago that we were expected to only speak publicly of respectable people and things. We still have the idiom 'if you have nothing good to say, say nothing at all'—a command that urges decorum and tact. As we will soon see, the speeches found in the Phaedrus are very tactful.

Perhaps a discussion of tact seems out of place in a paper which focuses on logic. Tact has to do with what is better left unsaid. Tact is a social grace, a skill, a key tenet of civility. It is also a rhetorical tool; but, it is no less necessary for dialectic. Tact is concerned with what is to be said as well as where and when. Certainly, tact has no place in analytical logic, but in speech, in *logos*, it is crucial. A proper speech ought to address everything that is relevant and forgo talk of anything superfluous. A good speech satiates our curiosity by giving a complete account of a given topic. It does this by including everything relevant while excluding everything which is irrelevant.

In the name of completeness, it is worth mentioning how completeness fares in our modern and ancient uses of *logos*. Completeness is a common goal of the science of inference was well as all sciences. It is the goal of any given science that it seeks a complete and unassailable account of the phenomenon which it studies. The suffix 'ology' is derived from '*logos*' and is applied to most scientific studies. So, for example, psychology seeks to give a complete and coherent account of the psyche, a complete *logos* of the soul. Above, Crowley and Hawhee state that *logos* is bound to ideas of personal history or reputation, but this concept of accounting holds the same sense of testifying to the whole truth and nothing but the truth that would apply equally to personal and impersonal topics. Furthermore, this 'accounting' also implies a gathering of one's experience in order to bring them into a discussion. Thus *logos* both refers to the process of expression as well as the requirement to express the whole topic.

Lastly, *'logos'* has a psychological dimension. I think this is most pressing when considering that *'logos'* may be translated as 'purpose'. 'Purpose' seems to neither be at home in colloquial English nor in logical theory. It is squarely psychological or teleological. Formal logic tries to keep to what can be derived from the data—rather than the use and direction of an argument, or the intent of the author. These linguistic elements better suit the study of rhetoric. Still, it's not completely foreign in colloquial English to question the 'logic' of an argument when we are really implying that the speaker is trying to dupe us-when we rhetorically ask just what the speaker is trying to achieve. For example, someone might ask 'just where are you going with this?' or, accusatively, 'I don't follow your logic!' Such phrases euphemistically seek to inquire about intentions rather than facts; these phrases are also somewhat anachronistic but I have heard them in use, and they will help to illustrates a move made in the Phaedrus that will be discussed later: that ends matter or—as I'd like to put it—that the telos of a speaker shapes their *logos*. The psychological ends shape the linguistic means.

Still, though a firmly psychological term 'purpose' has its analytical correlative as it implies non-contradiction; one who has contradictory beliefs is said to be deceived, confused or irrational. It could thus be argued that such a person couldn't be driven by purpose, but rather by error or insanity.[3] The weirdness of translating *'logos'* somewhat disappears when taken in context with another curious Greek word: *'ethos'*. *'Ethos'*, like *'logos'*, is another of those good Greek words which doesn't translate neatly. 'Ethics' comes from this word but morality has less to do with the original meaning than we moderns may assume. Rather *'ethos'* refers more so to character or personality. *Ethos* squarely pertains to the social standing of an individual or character, how revered or trustworthy one is; especially in reference to excellence in their chosen profession. Universal implications of moral rightness and wrongness (though important) pertain only tangentially.

6.3 A Country Setting

The *Phaedrus* has a somewhat novel setting. Its most pressing feature corresponds with its most noticeable deviation—from Plato's typical characterization of Socrates. Socrates is outside his typical haunt; he's outside of Athens. He is also out of his mind (or at least out of character). He has lost his composure and is, if not more verbose, more long-winded and more eloquent. This is not Socrates as we know him. He professes a malady; he is madly in love with giving and hearing speeches—a far cry from his typical insistence that he can neither make nor understand long speeches. He is also on his way to meet Phaedrus in the countryside. His madness has overridden his hatred for travel and Phaedrus is waiting with promises of fine words—Socrates' new true-love. In the wilderness Phaedrus is alone but not lonely. He has a scroll in hand that he has litigiously been mulling over. He tries to hide this fact but Socrates is aware of Phaedrus' proclivities. Phaedrus, like Socrates, is also in love with words. Thus

---

[3] Error and insanity are also major themes in the *Phaedrus*.

Socrates anticipates finding a lengthy and impressive scroll hidden under Phaedrus' toga. Phaedrus does not disappoint.

The scroll is authored by a respected wordsmith named Lysias—a rhetor who happens to have a special interest in Phaedrus. The topic of his speech is love. Specifically, it is about whether it is preferable for a young man to engage carnally with a man who does not love him as opposed to one who has genuine affection. When Socrates arrives Phaedrus can hardly contain his excitement—finally someone to share in his bacchic passion for words. Socrates is enraptured with the prospect of artful rhetoric so he hurriedly prompts Phaedrus to share his cherished scroll. Having read Lysias' speech Phaedrus waits with baited breath for Socrates to lavish it with praise. Instead Phaedrus is rebuffed and, after a bit of coaxing, Socrates makes a speech of his own.

Though recited by Socrates, this second speech is much like the first. Socrates takes the same position as Lysias. The premise remains unchanged. Socrates' speech is still about how it is better for a young lad to consort with men who do not love him, and to avoid admirers like a plague. Socrates goes into great lengths, chastising the devious insanity of lusty old men with culpable passions. His account is so thorough that his rhetoric besmirches the very nature of love—love is a sickness to be avoided at all costs. Phaedrus is quite impressed by Socrates' retort; but, just like with Lysias' speech, Socrates is still dissatisfied. In fact, he is even more perturbed in that he had carelessly blasphemed in his moment of erudite passion. He had been so disdainful of erotic love that he had insulted Eros, the god of physical love. Mortified, Socrates then fabricates a third speech, a Palinode, or apology to the gods.

Socrates' second speech, his Palinode, is much more measured than his first. He loses his vitriol for sexual attachment and replaces it with a very thoroughgoing exposition of what love is and how it benefits humanity. Love changes from being a malady to being a temporary mental distortion which ultimately leaves the psyche in a better condition; that is, after its initial discomfort and confusion. The Palinode is the most logical and organized speech of the three. But its order and logic may not be immediately apparent to the reader. It may seem convoluted rather than thorough. Its elegance will soon be explained as it is the *logos* of these three speeches—and the *ethos* of the speakers—which we will now explore.

6.4 The Opening Gambit: Lysias' Speech

The first speech (Lysias') begins by personally addressing Phaedrus. Or at least that is what we are led to believe. He opens by saying 'you understand my situation' (231A). Phaedrus' is never actually called by name. Neither author nor audience are ever personally addressed. Lover and beloved are spoken of, but only abstractly as Lysias initially appears to be squarely interested in coldly dissecting the logic of this erotic relationship. Yet, bit by bit, he becomes more and more familiar with his audience, with Phaedrus, while the subject matter correspondingly takes on a more personal character. True, some familiarity exists from the beginning; Lysias begins with the pronoun 'you' and consistently employs it. The barrier between the 'non-lover' and himself slowly erodes until 233B where all pretext of abstraction is abandoned; there

Lysias says 'Another point: you can expect to become a better person if you are won over by me, rather than by a lover'. Thus Lysias reveals that he isn't talking about 'non-lovers' in any abstract or absolute way; he is, instead, speaking directly of himself. This may be nothing. Perhaps Lysias is just being stylistically sloppy. However, in that his reputation precedes him, this inconsistency prompts us to believe that any bugs in his code are actually features. Read with suspicion, the encroaching familiarity of Lysias' style take on the character of a grooming technique. Is Lysias leading his audience to some predetermined end? Is this speech really written for Phaedrus personally? Or is it merely a clever piece of seduction meant for anyone like Phaedrus—a young, handsome man who is interested in the art of rhetoric? I would wager that it is, but, moreover that the ambiguity of Lysias' intent is intentional.

Looking more favorably at Lysias: his words tell us much about the social norms of intimacy during his time. His speech begins with a statement which implies a great deal of closeness. It's merely taken for granted that the audience, i.e., Phaedrus, is well versed in the argument which is about to be made. The speech, then, is something of a personal memory aid. One specifically crafted for Phaedrus. Thus the speech intentionally has no real introduction. This is to say that it doesn't, as Socrates will do, begin by defining terms. This neglect of an introduction would not be uncommon for the times. Socrates prides himself as a dialectician; Lysias as a rhetor. As rhetoric it is not bound to the same rules of style dictated by dialectic and so Lysias isn't required to begin with definitions. More importantly, in rhetoric it was (and still is) encouraged to tip-toe around subjects which were considered taboo or obscene—and this topic definitely meets that criteria. Sexual relations between men and teenage boys was much more accepted in Lysias' time but it still was not an honorable pursuit. Thus Lysias, in neglecting to define the detailed terms of a disreputable relationship, displays his tact and understanding of what is considered fit for proper public Athenian discourse. In a way, Lysias gives the subject matter its proper due by foregoing an introduction in favor of a hushed beginning which assumes prior understanding. Thus, by leaving out what ought to go unsaid, Lysias displays that he is a shrewd craftsman, a skilled sophist.

Interpreted suspiciously or not, the speech is a hybrid of formal and informal style: it masquerades as a sort of objective treatise on erotic love but it soon loses its objectivity. It shifts from speaking in the third, to the second person and so has a semi-authoritative tone. Again, this stylistic quirk could be a feature rather than a flaw. Lysias' seduction is built on the preliminary construction of an ideal relationship (i.e., that non-lovers make the best mates). This is followed by the unspoken recommendation of its adoption (that Phaedrus should cede to the very dispassionate Lysias). How fitting, then, that his style conforms to his content. Lysias effortlessly shifts from the descriptive to normative without declaration. Simultaneously he slips from an objective, formal tone, to an informal one in a sort of sexy sleight of hand of which Phaedrus may or may not be aware.[4] This transition from the formal 'lover/beloved' to the informal 'I/you' relationship may be a strength of the speech (as it is rhetorically clever) or it may be the crack which subtly suggests a darkly deceptive side of Lysias. This turn from the

---

[4] It's a distinct possibility that Phaedrus is fully aware of Lysias' tricks and doesn't care. It's possible that Phaedrus approves of Lysias' persuasion techniques. In other words: he could be into it.

6.4. THE OPENING GAMBIT: LYSIAS' SPEECH

distant to the familiar may suggest rhetorical mastery, or it may suggest a predator/prey relationship. Likely it is meant to express both.

After his terse beginning, Lysias commences his argument with an interesting metaphor. He likens a lover to a miser. This is the first of two money metaphors which bookend his speech. We are told that a lover 'keeps his eye on the balance sheet-where his interests have suffered from love and where they have done well' whereas a non-lover conducts all business sober-mindedly (231B). Much can be made of this metaphor. It foreshows what Socrates will soon allude to—that Lysias is exactly this sort of miserly lover which Lysias himself condemns. But Lysias has his own purposes in mind. He is suggesting that the lover does not have his own house in order. The lover is anti-social; he is more concerned with his ledger than the community and thus is apt to blame outside forces for his own shortcomings. From here forward praise and blame are to be the basis that Lysias uses to portray the lover and the non-lover.

As for praise, the lover divides loyalties through his fickle alliances. He praises one boy before moving onto the next. Once separated, the lover transfers his kind words to his new love and leaves the last with reproach. He cares only for himself and leaves discord in his wake. Yet, he ultimately hangs himself with his own words. Lovers are apt to self-report love-sickness and use it to excuse their horrid acts. Shrewdly, Lysias uses this confession to ask if it would even be right to hold a sick man to the words spoken under duress, of unsound mind. He leads us to the magnanimous answer of 'no', as it only makes sense that a contract made without psychological soundness is devoid of meaning. We are to pity the madness of the love-inflicted; and, though such a cad can't be held culpable, the best and most noble action is to avoid such a rake.

Thus, praise and blame lead us to consider social participation—who to shun and whose company to keep. Lysias notes that the healthy stock of non-lovers will give a young lad a much larger pool of admirers. This isn't (we're assured) to laud popularity contests but rather to ensure that the boy has a better chance of finding a fine suitor. On the contrary, the lover will be boastful and 'seek the glory that comes from popular reputation' (232A).[5] He'll also be jealous of others; he'll isolate the young man. He will assume that his beloved will be as flippant as himself; and, just as the lover will jump from beloved to beloved, so too the beloved won't hesitate to find a better lover. Thus, the wages of jealousy are borne out by the beloved. Whereas the boy begins the relationship hoping to become socialized into greater society the opposite occurs and all parties are harmed. Except, perhaps, the lover. It seems implicit that the lover will profit at the expense of the community, but this isn't explicit. Socrates will be more explicit in his accusation that the lover harms all, even himself. Contrary to a lover, a non-lover will be civil and cordially introduce the youth into a larger sphere of influence and social standing. Likewise, he will also love and care for the boy's soul rather than the other ravenous, unruly base admirers who are like pitiable beggars with their hands outstretched.

The beggar metaphor is the second bookend to the speech (233E). Lysias hints that, like beggars, lovers will give you nothing in return but a bit of gratitude, which

---

[5] I surmise this point is meant to suggest that a miser's hoard of gold is like a demagogue's hoard of followers.

is likely to be as fleeting as their full bellies. As soon as they are hungry again their gratitude will evaporate. Perhaps they will even, in their meanness, resent the initial act of charity. The two metaphors aren't merely for decoration. The suit begins with a metaphor concerning wealth and ends with a another. As a pair they represent two sides to the same coin.[6] The miser harms society by being withholding, the beggar harms the group by being prodigal. The mean of these two vices is the virtue that the non-lover is meant to resemble—a man both liberal in money and love. This is the mild character that Lysias means to project upon his audience.

A final note on style: the strengths of Lysias' speech rest upon a likely story—a hypothetical construct of Athenian stereotypes. I don't say this to denigrate the piece. Stereotypes come to us for reasons—they mirror popular opinion. Furthermore Lysias' speech is built on *enthymemes*, which are something like stereotypes. The word 'enthymeme' comes from '*thymos*' the Greek word for 'spirit'. Thus an enthymeme captures the spirit rather than the essence of things.[7] Lysias' portrayal is likely close enough to the mark to illustrate a plausible scenario and plausibility is useful when persuasion is your goal.[8] However, plausibility isn't enough for Socrates, who is disdainful of such likely stories.[9] Socrates prefers exactitude—the essences of things, or things as they truly are—but this dialectical love of definitiveness is less important to the art of rhetoric. The stringent demands of Socrates are his own and what Lysias is doing would be considered par for the course and artfully done. Thus, Lysias' speech, by demonstrating his rhetorical ability, is a perfect example of a rhetorical 'argument from *ethos*' (Crowley and Hawhee, 83).

Lysias' true character will be discussed later, however; for now, let's recap what we know for sure: Lysias wishes to have sex with Phaedrus; he wishes to explain why; and his suit rests in the argument that one in Phaedrus' situation should accept the advances of one in Lysias' position (i.e., one who does not love him rather than one who does). Roughly put, the thrust of Lysias' rhetoric is simple, he is a cool guy who can keep his illicit affairs under wraps while being a great mentor. According to him the rest of Phaedrus' suitors are worse than buffoons. They are irresponsible men who are beside themselves with passion and who cannot be trusted. If this pitch weren't enough, Lysias also displays how adept he is in the art of rhetoric, the very skill which has wholly captured Phaedrus' interests. This demonstration of artistry is the unspoken cherry on the top of his gambit and what characterizes it as an argument from *ethos*. Lysias demonstrates his trustworthiness by expertly assessing Athenian civic life, his role in it, and his ability for verbal mastery.

Lysias is not creating an *ethos*, or character, out of whole cloth; instead, he pieces together the bad behaviors that love-maddened suitors are thought to display. He skillfully uses euphemism and enthymeme to display his social tact and charm the very

---

[6] Pun intended.

[7] We have a very modern equivalent to this. We talk about the 'gist' of things which comes to us from the German word 'geist' which literally means 'ghost' or 'spirit'.

[8] It should also be noted that speaking in enthymemes would be par for the course when speaking of unseemly things (p. 157).

[9] Or at least he thinks plausibility is valid only for certain explanations; and, given the tone of the Phaedrus, rhetoric ought not rely on likely stories.

spirited Phaedrus by appealing to his good character. Primarily, the speech is an expert show piece, an *epideictic* (Nehamas and Woodruff, xvii). An *epideictic* is designed to express the rhetorical mastery by persuasively arguing wild claims (Socrates would likely refer to this as making the weaker argument the stronger). Fitting then that Lysias topic is so ambitious—even paradoxical.[10] One would think that a non-lover would have no interest in love. The fact that Lysias can play with the definition so persuasively shows just how good a rhetor he is. With this in mind we can see how Lysias tries to both simultaneously display his worldliness in the matters of love while displaying his aptitude for rhetoric. All this is done while he also indirectly contrasts himself to a very ugly persona. Thus, he establishes an argument from ethics by displaying his temperance in matters of love while also showing off his chops. Looked at this way one can see how it is a very successful ethical argument. Not only does it demonstrate technical mastery but it also establishes trust, which is crucial for persuasion.

But success is nothing without a criterion. Persuasiveness is the rhetorical criterion for success and Phaedrus is the sole arbiter in this case. The purpose of Lysias' speech is to bed Phaedrus. Hence we must, by Phaedrus' enthusiasm, say that it is partially successful. But only partially in that Phaedrus is still deliberating. Psychologically, he seems to be sold, but he hasn't yet made a purchase—and that is the true measure of success. In a sense the speech paints a good likeness of his current state of affairs. Thus, Lysias certainly gives a good account of himself and Athenian life as it pertains to sexual relations. The plausibility of his account is reflected in Phaedrus acceptance. This is revealed when he gushes to Socrates, asking, 'Do you think that any other Greek could say anything more impressive or more *complete* on the same subject?' (emphasis added, 234E) Phaedrus expects a resounding 'yes'. Instead he is met with a measured agreement: impressive, yes; complete, no.

6.5 Phaedrus rebuffed: Socrates' First Speech

Interestingly, Socrates doesn't take Lysias to task by analytically dissecting his speech—he is far too playful (and too shrewd) for that. Instead he insists he's already heard a better speech on the same topic. Phaedrus is incredulous, he thinks that Lysias has given a complete account and nothing more can be added either in style or content. So, would Phaedrus be persuaded by a dryly analytic, methodical analysis? Unlikely. Phaedrus is wowed by the ethical argument, he is captivated by artistic mastery. He loves how fantastic the claim is and is also likely frightened by the idea of being involved with such a disgraceful character from which Lysias distinguishes himself. Socrates knows Phaedrus' character and he knows that Phaedrus will only be moved by a greater show of technical rhetorical force. Thus, Socrates has to craft an even better argument from *ethos*—he needs to beat Lysias at his own game.

Socrates begins his retort after some playful preamble. Much of this can be glossed over for our purposes but it's important to note that Socrates does not begin until he establishes that his argument is to use the premise that 'the lover is less sane than the non-lover' (236B); and that though Socrates is unable to make or understand long

---

[10] In the most literal sense, *para-doxa*, against (common) opinion.

speeches, he thinks that he can recall one that very conveniently happens to fit perfectly with the topic at hand. Also it just so happens that this speech begins by saying:

> There once was a boy, a youth rather, and he was very beautiful, and had very many lovers. One of them was wily and had persuaded him that he was not in love, though he loved the lad no less than the others. And once in pressing his suit to him, he tried to persuade him that he ought to give his favors to a man who did not love him rather than to one who did. And this is what he said... (237B)

This detached air directly mocks Lysias' style while laying his intentions bare. Socrates is clearly creating a parody of Lysias; but, as we will see, he makes some very important distinctions. Socrates is making an argument from ethics but he is doing it philosopher-style. Though Socrates, in a sense, becomes possessed by Lysias he still maintains his Socratic penchant for rigor. And so he sets out to best Lysias by making Lysias' own argument more logical and methodical. Lysias built his argument on a kind of Frankenstein lover—a piecemeal collection of bad behaviors that aren't beyond the realm of reality. By contrast, Socrates proceeds by the hypothesis that the lover is less sane than the non-lover. There are three important terms in this phrase: love, sanity, and the relationship of greater/lesser. These were the same basic ideas that Lysias used however they weren't expressed with any logical or philosophical rigor. Lysias never asks what 'love' or 'madness' truly are. Rather he simply asserts that the lover is less reasonable by way of a few invented examples.

Lysias begins his assault with a metaphor of how the lover is like a miser. Socrates mirrors this by likening the lover to a glutton. His style, of course is more exacting. He begins by establishing that all men, even non-lovers, are naturally drawn to beauty; that all men are born with an innate drive both for pleasure and excellence; and that these urges have the power to override the powers of reason in a tyrannical fashion (238B). Mindless indulgence in food is called 'gluttony'; likewise a mad feast of passion is called 'eros'. Socrates' lover is like Lysias'; he is vicious, but this time he is so by logical necessity rather than conjecture. In Socrates' tale, the tyranny of pleasure causes the lover to work against the boy's best interest in a *systematically* perverse manner. Simply put: the lover is *pathologically* evil. Socrates' methodical start forces him to bypass the lover as Frankenstein monster and leads him to something completely devilish. And so Socrates' lover doesn't just look for an attractive victim, but also one who is vulnerable. His version of the lover wants an easy mark so he targets the poor and orphaned. He prays on the weak, he seeks inferiors so that they can be more easily controlled. He is unambiguously predatory.

Lysias' tale of the lover focuses on the harm done to society. According to him, the lover causes strife through disloyalty. He hides his beloved for fear that the beloved will find more worthy men. Socrates' lover shares the same behaviors but his internal psychology is much more twisted. Instead of being jealous of others the lover is jealous of his beloved and keeps him away from others for fear that they may improve and strengthen him. The lover dreads the day his beloved might outgrow him. He fears that others may give his sex-slave enough strength to leave his abusive situation. This is a

6.5. PHAEDRUS REBUFFED: SOCRATES' FIRST SPEECH

subtle but important distinction. Socrates is taking depravity to the extreme in order to show that absolutely no good can come of it. The lover does his level best to keep the boy weak and helpless. Thus, not only does he attempt to extinguish any virtue in the boy, but he also discourages him to develop physically. He also lavishes his beloved in flattery (just as Lysias' did) but—instead of holding his tongue until he has a new victim in hand—Socrates' lover loosens his tongue with drink and lavishly berates the boy. Socrates' lover is completely incontinent. He is an absolute brute—he's more animal than man. Socrates hammers this point home when he says 'Do wolves love lambs? That's how lovers befriend a boy!' (241D). Thus Socrates hamfistedly alludes to the predatory nature that was hidden in Lysias' speech.

And then nothing. Socrates, about to sing the praises of the non-lover, stops dead and dismissively says that '[...] in a word, every shortcoming for which we blamed the lover has its contrary advantage, and the non-lover possesses it' (241E). Why? According to the text Socrates realizes that Eros is divine and thus a lover (i.e., someone inspired by Eros) could not act so brutishly. I believe this is said in earnest but that there are also other, more mundane, reasons that Socrates leaves the ways of the non-lover unspoken. First, Socrates is effectively divulging the magicians' secrets by trivializing the exercise of rhetoric. He's basically saying that there's nothing special about rhetoric once the steps are laid bare, when the logic, or underlying mechanics, of the argument is understood. At this point it is child's play, it's something like a paint-by-numbers exercise. Yes, Socrates is worried about blaspheming but, more importantly, he knows that if he were to spell out how heroic the non-lover is he would be forced (by logical necessity) to describe how such a character would seek nothing from his beloved. Nothing, not even sex.

Further (perhaps more importantly) it may also be that a non-lover is an impossibility. Socrates finds no challenge in describing someone who is like a pure Freudian Id, but to describe one completely devoid of eros is altogether different. Perhaps Socrates is suggesting that a person without an Id may be a logical impossibility; something like a category error. Socrates has already established that all men are of a particular kind that are drawn to beauty. Without Eros there would be no motivational force. Such a creature would be without purpose and thus would be absurd in a very serious way. Some or all of these possibilities may be true; which ones makes little difference. Most importantly, had Socrates continued his speech the jig would be up as he would violate *his original aims*. The official purpose of this speech is to demonstrate why it is good for a boy to cede to a non-lover rather than a lover; yet, if the perfect lover is celibate then the purpose of inquiry loses its point. Sexuality is inherent in the premise, yet it seems like the conclusion will demand Platonic love. Clearly something isn't quite right with this argument. Socrates may blaspheme if he proceeds, but he may also lose the game and, most importantly, Phaedrus' interest.

Socrates' first speech is literally incomplete, but does it give *a good account*? It certainly seems more logical and psychological to our modern ears. As far as rhetorical success, though incomplete, it seems (like Lysias') to have partially met its mark. Phaedrus, if not convinced, is at least disappointed that Socrates refuses to go on. He is also gluttonously delighted that Socrates now wishes to make a yet another speech, a Palinode, an apology to the gods. In Socrates' first speech he plays with the extremes that Lysias loosely provides. The logic of Socrates leaves enthymemes in favor of pathological behavior; and, by proceeding in more defined terms he takes Lysias' argument to its logical conclusion where he nearly falls into contradiction. Thus by constructing his Palinode, Socrates will attempt to steer away from conflict by founding a new beginning.

Socrates second speech preserves the premise of his first. He still hasn't abandoned the presumption that 'the lover is less sane than the non-lover'. Socrates' first speech squarely targeted the contrary and binary relationship of the lover and the non-lover. His second speech redefines the other two terms in his thesis: 'love' and 'madness'. Rather than immediately pitting lovers against non-lovers Socrates starts by defining terms. It should be noted that this is the first speech which contains a true introduction. Socrates begins by defining the various kinds of madness. He speaks of diviners and oracles and even lumps philosophers and lovers in this same group. Such people are mad in a beneficial way. These people aren't simply crazy, they are divinely mad (*theia mania*). They have divine knowledge yet only appear insane or irrational. Kinds of lovers are also distinguished. Admittedly, less time is spent on distinguishing kinds of lovers than kinds of divine madness but this fits as Socrates' main goal is still to describe the ideal lover rather than its variations. He also spends little time on what poor lovers are like (but we've heard plenty about that already in his first speech).

Socrates does mention that followers of Aries can suddenly turn 'murderous' (252C). But he doesn't outright chastise them. I suspect that he does this to create a complex schema which better represents the various psychologies of men. Men can be more martial or more fair and, I think that Plato has his preferences, but he doesn't let them color the dialogue. Instead he reintroduces his theory of conflicting psychological drives with the analogy of the charioteer. The soul as charioteer has three parts: the charioteer and his two horses. One good horse represents the desire for virtue, while the other symbolizes the carnal urges. This rehash of the first theory gives reason a more prominent place as the charioteer. The charioteer is the one who pursues the good and the beautiful. He is aided by the good horse who 'is guided by words alone' (253E) and struggles against the bad (who needs constant whipping).

The paradigm that Socrates adopts in his second speech is essentially the same as his first. Thus it comes as no surprise that Socrates' new imagining of the lover is exactly as we would predict. Like clockwork, the absent (unspoken) 'non-lover' from Socrates first speech appears in his second redubbed 'the lover' (celibacy and all!). Why does Socrates do this?—he does this to demonstrate how rhetoric can make the weaker argument the stronger. The first two speeches were contests premised on the goal of making an outrageous, absurd claim palpable.[11] Initially, Socrates treats this as a fun game but not one to be taken seriously. Phaedrus convinces him to take it up in earnest and he eventually finds that he's been given an impossible task—or at least this task is impossible given Socrates' impeccable character as this rhetorical challenge would have led Socrates to impiety and absurdity.

---

[11] I believe that by refusing to finish this speech Plato is leaving the absurd unspoken, playing with the meaning of '*alogos*'. He wishes to illustrate that one can play with absurdity, but eventually you'll be forced to talk nonsense.

Character is a very important part of the dialogue and Plato is as wily as Lysias on this issue. The dialogue is infused with veiled purposes, subterfuge and allusion. Lysias portrays himself as honest and straightforward-someone very worthy of consorting with. Socrates lays Lysias' intentions bare and insinuates that he is the worst kind of snake in the grass. And yet Socrates doesn't claim to be making this allusion. Instead he insists he is possessed. In his manic state he just happens to sound a lot like Lysias. And yet, Socrates takes responsibility for his possession when he realizes that it is verging on impiety. This is somewhat ironic given that both he and Lysias have already established that one isn't completely culpable for their actions when overcome by passion. So what does Socrates do when he comes to his senses? Does he simply say he's sorry and walk away? No, his solution is to stay and, yet again, become possessed by a different spirit so that he can make an apology for the first!

6.6 Socrates' Second Speech: The Janus-Faced

This is utterly schizophrenic. What is Plato doing? In typical Platonic fashion he's doing many things at once. On the surface he's simply being consistent with Socrates' final conclusion: that getting carried away is better than miserly self-control (244B). But more importantly he's illustrating the multiple *logoi* of the charioteer. Plato's psychological paradigm splits the soul into three in order to simplify it, but the soul is never such a static perfection. Rather each character has their own charioteer and each have their own logic just as they have their own motivating force. Lysias roughly represents the bad horse, Phaedrus the good (the one guided only by words) and Socrates is the charioteer; more precisely, Lysias is *dominated* by his own bad horse, Phaedrus by the good and Socrates is the only one with a modicum of control.

Socrates uncovers a more sinister view of Lysias. Lysias is an old man who invests his time seducing the young. His seduction is defeated by Socrates, who implies that Lysias is not who he says he is. But does Lysias *think* he is as he describes? Is he lying or deluded?—likely he's delusional, but we can't know for sure. Lysias' absence means that his true intents can never shed their ambiguity.[12] Still, it was established that when the erotic drive is left unchecked it rules over reason like a tyrant, and Lysias is lusty. This renders the self-knowledge and self-possession of Lysias a moot point. Thus Lysias' seduction piece is perhaps as much a deception as it is a delusion. His logic, his account or assessment of the situation, is likely as complete as he can give. It is also as complete as the impetuous Phaedrus can imagine. But it is partial. In fact, it is worse than partial, it is confused.

Lysias' speech is unruly and disordered, which is indicative of his unruly and disordered mind. He can't think straight! His reason is overcome! He *is* absurd! And so his suit of Phaedrus bears this out in a kind of aphasia. His speech is disordered in every possible way: it lacks a logical flow, it lacks a thesis and (perhaps most importantly) it advocates something improper. Still, his account anticipates Phaedrus' noble desires and thus it gropes at the truth. Phaedrus, driven by a desire for social status, is susceptible

---

[12] The dialogue remains consistent in that Socrates establishes that the necessary absence of the author is a defect of writing as a medium of communication.

to praise and promised glory. He loves the glory of winning and he wants to be lauded like the great champion, Lysias. Shrewdly then, Lysias' promises greatness and this intrigues Phaedrus. So Lysias writes with this in mind, emphasizing how his bargain can lead to greater social harmony and respect for all parties. His purpose guides him through nobility back to his ignoble appetite, and Phaedrus is with him every step of the way.

Phaedrus is originally convinced of Lysias' good character but Socrates subtly casts doubts on this perception. So what is Lysias truly like?—It may be that Lysias is as grotesque as the monster depicted in Socrates' first speech. It could be that Lysias is an accomplished liar. Or maybe he's simply struggling to understand the nobility of Phaedrus. Perhaps a part of Lysias loves Phaedrus' honor-hungry youthfulness. Perhaps he has some real, though perverse, paternal feelings. Perhaps his feelings are simple and honest. Perhaps his best advice to Phaedrus is a simple description of how to get ahead in contemporary Athens. These possibilities, though worthy of exploration, are tangential to our inquiry. What matters is that Lysias' *logos*, his account, is truthful yet lacking. It is incomplete and perhaps incoherent[13] but it is truthful in that it reveals Lysias' *ethos* and that his *logos* (i.e., purpose/reason) is driven by appetite. His *logos* is indicative of his *ethos*.[14]

As for the second speech, we ought to take Socrates at his word—he is possessed by an alien mind, or at least a character not his own. He's just drunk on words. The logic he employs, the accounts he gives, are not wholly his own. His speech is false in that it doesn't reveal his character yet true in that it reveals a character. He speaks as though he is channeling Lysias but his persona is actually mimicking Phaedrus'.[15] The purpose of this speech (aside from embarrassing Lysias) is to win for winning's sake and thus the product is vain-glory. But vain-glory *is* Phaedrus! Or more charitably: he is a spirited youth who cannot distinguish between glory and vain-glory. This is why Socrates begins his first speech by shaming Lysias-Phaedrus, driven by words, is highly susceptible to shame.[16] This is precisely the tact that Socrates needs to take in order to convince Phaedrus. But had Socrates finished his challenge he would have also shamed Phaedrus, for Phaedrus has already wedded his pride to Lysias' composition, and pride is a good more dear to Phaedrus than his own body. In allowing Phaedrus to save face Socrates cements his own persuasiveness. So, just as Lysias' account revealed his character and that that particular character was ultimately shaped by a particular reason so too is Socrates' first speech revelatory. Socrates' original speech reveals Phaedrus' *ethos*: a character driven by glory.

---

[13] At least when transliterated in Socrates' first speech.

[14] Thus we see why Lysias' speech begins and ends in wealth metaphors. He is revealed to be akin to a money-lover. He thinks in terms only of coin. His denunciation of the money-lover is either a smokescreen (a deceitful projection of himself onto others), a sign of some mental disorder or some other confusion.

[15] Socrates alludes to this fact at 242D when speaking of the divinity of Eros: 'Well, Lysias certainly doesn't [believe] and neither does your speech, which you charmed me through your potion into delivering myself.' Before hammering the point home in 244A where he says 'You'll have to understand, beautiful boy, that the previous speech was by Phaedrus [...]'.

[16] Susceptibility to shame or 'shamefastness' was considered a virtue in the Greek world, and it is a virtue which Aristotle described as strongest among the youth.

6.6. SOCRATES' SECOND SPEECH: THE JANUS-FACED

Perhaps out of humility, perhaps, again, to save face, Socrates shirks authorship of his Palinode. Remember: Socrates insists that he can neither make nor understand long speeches. Thus, at the very least, Plato must put words in Socrates mouth as a literary device. One which serves to keep Socrates' behavior consistent. But I think there is a larger point here. Extremes are commonplace in this dialogue and reality is presumably found in between. Socrates is something of an avatar for philosophy but this doesn't mean that he can speak from the seat of divine knowledge. During the preamble, before his second speech, Socrates asks Phaedrus if he noticed how he has 'passed lyric into epic poetry' (241E). This may connote both that the kind of depraved individual formally described as a 'lover' is a rare beast and also that his opposite is an equally rare heroic type. One whose deeds lend themselves to epic verse. Perhaps, then, the lover is something gallantly half-divine.[17]

Socrates is no hero, not in the ancient Greek sense of the word. But luckily for us he is an empty vessel, waiting to be filled. Once filled, Socrates becomes the clear-sighted charioteer. He sees love and madness for what they are. He also sees clearly how the soul is divided and other prophetic visions. Perhaps Socrates is being coy. Perhaps he really is representing himself in this speech. Perhaps he's representing the author, Plato. Or perhaps he's simply dramatically playing into the other extreme of the topic at hand—the perfect lover, the perfect philosopher, a sort of mad oracle. Whatever the case for possession this persona is squarely in pursuit of the true the good and the beautiful and his character is meant to be something truly glorious. In this way we have a third demonstration of a kind of *logos* and the *ethos* which it embodies. This final composition, this Palinode is, in a sense, divine. It is beautiful in its purposes but it is also beautiful in its method. The three speeches ascend from less to more logical in that they begin disjointed and meandering and become more well spoken and pointed—so too do the characters which speak.

And so we have it. The many logics of the Phaedrus (and their corresponding ethics). I have attempted to disentangle the many senses of the word that Plato has carefully woven into the dialogue. I assume I've missed much but I hope I have given the reader a primer on how to approach some of Plato's eccentric literary devices, a glimpse of how he views the human soul and where he places logic, or reason, in relation to it. There is an objective hierarchy within the soul which proceeds from the rational, through the spirited and ending in the appetitive. Likewise this hierarchy mirrors society and its citizens. Lysias speaks from the appetitive part of his soul as his judgement is clouded by its dominance. His purpose permeates his discourse in spite of himself; and, as a result, his account of love is partial to his purpose. What it fails to gather is accounted for in Socrates' second speech (or is it Phaedrus'). Socrates' supplies the missing piece by suggesting that a truly honorable and magnanimous person gives of himself without asking for anything in return. The only glaringly confused part of this speech has less to do with its content and more to do with the words employed in its exposition. It's blindly noble in its pursuit and so its accounting of love blunders about in a sort of

---

[17] At 244A Socrates attributes his second speech to 'Stesichorus, Euphemus' son, from Himera', which translates to 'Stesichorus son of Good Speaker, from the Land of Desire' (Nehamas and Woodruff, 27). The convenient title of Stesichorus suggests to me what he is not meant to be taken as a real individual. But this is certainly an alternative possibility.

autistic fashion. It barrels towards truth with blinders on and so it unknowingly tramples it by making the weaker argument the stronger. Finally, Socrates' Palinode restores order by setting the record straight, by putting 'the lover' back in its rightful place as one who is truly honorable and magnificent.

6.7 *Logos* and *Alogos*

The direct translation of *logos* may be logic. Would, then, *alogos* be illogic? Perhaps not. *Alogos* in modern Greek translates to 'absurd' or 'irrational'. But just as the ancient use of the term refers more to speaking, perhaps *alogos* may better fit as what is left unsaid. The *Phaedrus* is a book which builds dramatic irony by leaving much unsaid. As mentioned, Socrates first speech leaves us wanting. Ironically this very speech is prefaced with the assumption that Lysias has figuratively left something unspoken but it is Socrates who is literally at a loss. Lysias has left out an introduction, he hasn't defined his terms; but, moreover he has hidden himself. True, he confesses his carnal interests but if he were truly passionless then he would have no physical interest in Phaedrus. Thus his claim of objectivity in the matters of love is a sort of psychological absurdity. If he were truly the 'non-lover', then his lack of passion would contradict his stated purpose: to have Phaedrus as his lover. But Lysias' absurdities don't end there. If the piece is an epideictic then it can be assumed to make paradoxical claims. In this case Lysias has inverted what it means to be a lover. His lover is actually a non-lover and his non-lover is truly in love (or at least lust). Thus his argument is built on a semantic absurdity, an inversion of terms. Lastly, his unstated premise is false, or at least it takes for granted that love is madness and that madness is necessarily evil. This last point is never expressed. It is the unsaid error which Socrates will eventually expose and correct.

Then there is the matter of tact; or what should and should not be said. Lysias displays tact by beating around the bush. He neglects to honor the indecent act he is proposing. Yet, when Socrates takes his argument to its logical conclusion he ends up speaking ill of the gods. Socrates, too, avoids getting into the grizzly details of pederasty; yet, in doing so he says what he ought not to say. He dishonors Eros by uttering slanderous rubbish. Not only is this an error in tact, but it is an act of impiety (another crime of utterance). On top of this, Socrates takes it as a bare fact that Eros is perfect. Thus, to speak otherwise is an ontological absurdity. Is Socrates' first speech more or less logical? Is it partially absurd? True, it is more coherent and better organized than Lysias', but it also commits the same inherent and fundamental errors. On top of this, it is blatantly incomplete in that it is left unfinished. Still, it may be a little bit pregnant. Socrates' first speech makes way for his Palinode, it shines light on Lysias' false premise and clears the ground for a better beginning.

It's debatable if Socrates' two speeches are, in fact, two. True, Socrates begins again, this time with defined terms and a modified premise. By giving a proper introduction he is forced to evaluate the nature of madness and includes it in his account; thusly, he gathers all relevant and previously neglected information. By doing so he honors his subject matter: Eros. He is also forced to account for a complex psychology. But is his Palinode complete in terms of the lover and the non-lover? Interestingly, Socrates' palinode doesn't mention the beastly lech that he and Lysias have been warning against. Perhaps enough has been said of him already; or, perhaps he is so below the level of civil discourse. Perhaps talk of the non-lover, then, would detract from the completeness of the Palinode by adding something superfluous. Why mar the beauty of the palinode by mentioning this beast in the same breath as the god? That would be tactless and better left unsaid. After all, if you have nothing good to say, it's best to say nothing at all.

6.8 Conclusion

I'll leave to the reader the larger questions. I've attempted to give the reader a primer on how to approach the *Phaedrus*, at least insofar as the three major speeches are concerned. This is really just scratching the surface. The three speeches comprise only the first half of the *Phaedrus*; and, though *logos* is primarily about the spoken word it can also mean 'writing'. This may yield some interesting results inasmuch as the *Phaedrus*' second half discusses writing in a negative light, as mere rhetoric, as dead words. Furthermore, not knowing ancient Greek, my linguistic analysis is limited to secondary sources. I've consulted philologists on how *logos* may have been used, but I would need to understand the original Greek to go into any further depth on the exact word play employed by Plato. Still, I hope I have established how Plato's conception of *logos*, though not as mathematically inclined as our own, is extremely complex and nuanced. I suspect that Plato's conception of '*logos*' is even more elaborate than I have expressed. Though, in the same token, I get the same feeling when modern logicians broach the subject of logic *per se*.

Personally I think that the *Phaedrus* is deeply ironic and humorous. Yes Socrates speaks ill of the written word, but I don't think that Plato completely agreed. Rather that he considered both written and spoken rhetoric to be more about intention and less about the medium of expression. Thus we see how entwined *logos* is with character and how character is to be analysed in psychological terms. I've tried to focus on how Plato's *logos* may be familiar to us. Fundamentally, '*logos*' is more akin to narrative and storytelling but it is not without formal elements. Purpose lends itself to purely analytical concepts like non-contradiction. Demands for a complete narrative leads us to ask if all parts of an argument are coherent or harmonious. Completeness also leads us to ask if any part of the argument is missing, if our terms are accurate, and if the argument corresponds with factual reality.

I think the most important feature of '*logos*' is how it makes intuition explicit. *Logos* turns thought into speech; which, in turn, tables ideas and makes them available for scrutiny. One can see how this would be important for Socrates' quest for self-knowledge. On the other hand, I think that logic is, in typical Greek fashion, more of a contact sport. I would say that Socrates' underlying objection to writing isn't that it is inherently inferior so much as that it diminishes the value of discourse, of dialogue. I think Socrates would object less to the written works of modern logicians and more to the solitary nature of the work. I think this is what he's getting at when he refuses to finish his first speech. He considers the exercise a kind of child's game—something (like sudoku) fun, but not to be taken too seriously. I don't wholeheartedly agree, and I doubt that Plato did either; but this belittling of abstract analysis is in line with our favorite gadfly and his quest for more practical wisdom.

The social aspect of *logos* is an interesting one. Socrates complains profusely about sophists and their ability to make the weaker argument the stronger. Lysias is a master in this regard. Yet, Socrates doesn't grandstand in his condemnation. Though he clearly disapproves, his condemnation doesn't take the form of a call for punishment. Socrates has the temperament of a doctor rather than an authoritarian. And though Socrates coyly navigates the boundaries of the discussion, he is more concerned with the psyche than with words. Words inform and inspire the soul. Thus, Socrates describes them as a potion. He imbibes them so much that they carry him away in absurd directions. He is possessed by them or is drunk on them—at least he pretends to be in order to illustrate the danger of admiring wordcraft above soulcraft.

Plato's '*logos*' is deeply tangled in notions of ethics and psychology. Is Plato's schema merely anachronistic or is there something to be said about how purpose shapes mental, linguistic or even logical habits? The idea of uncovering an *ethos* from an argument is hermeneutical and interpretation seems quite far from our current science of inference. But is there any room in this discipline for such concerns? Perhaps one could give Lysias more leeway and argue that his argument is logical in a paraconsistent way. After all, paraconsistent logic operates on the assumption that it is possible to reason using inconsistent information. And, as I've established, it's very possible that he lacks the self-knowledge to see the inconsistencies in his own argument. Or perhaps something completely different.

6.9 Bibliography

Braun, Eva T. H., (2011). *The Logos of Heraclitus: The First Philosopher of the West on Its Most Interesting Term.*, Philadelphia: Paul Dry, 2011. Kindle Ed.

Crowley, Sharon, and Debra Hawhee, (1999). *Ancient Rhetorics for Contemporary Students.*, Boston: Allyn and Bacon.

Grube, G. M. A. (Tr.) (1981). *Plato: Five Dialogues.*, Indianapolis: Hackett Pub.

Nehamas, A. and P. Woodruff (Tr.) (1995). *Plato: Phaedrus. Translated, with Introduction and Notes.*, Indianapolis: Hackett Publishing Co.

Part II

Papers Related to Schotch's Work

Chapter 7

Scotch on Quine, with Help From Carnap and Prior

M. J. Creswell[1]

This essay is in two parts. In Part I, I focus on examining the discussion of Quine's attitude to propositional modal logic in Chapter 4 of Peter Schotch's *Essays on Philosophical Logic*. Much of what I say here is suggested by an example presented in Prior 1964, which I use to shew that the semantics of the truth-functional operators of the classical propositional calculus is more closely involved with modality than many have thought. Part II takes a close look at the use that Prior himself made of this example in the article in question, since it suggests that, unlike Quine and perhaps Schotch, Prior's view of truth, and of its role in semantics is somewhat more sceptical.

7.1 Truth Functions and Necessity

Chapter 4 of Schotch (forthcoming) is entitled 'Quine on modal logic: Reimagined'. Schotch does not here discuss Quine's views on modal predicate logic, and I think he is wise not to do so, since the latter topic raises philosophical issues of a quite distinct kind from the issues Schotch and I wish to focus on. Schotch's main theme is that Quine's worries about modal logic are better put by admitting the distinction between language and metalanguage. Given Quine's belief that modal logic involves a use/mention confusion it is perhaps a little odd that Quine did not rely more on this distinction. At any rate, 'reimagining' Quine, using the distinction, seems to me the right way to go.[2] My comments will concern the discussion in Schotch's section 4.3, 'Quine *impensé*', specifically the material on pp. 32-34, in which Schotch discusses Quine's claim that, unlike the standard truth-functional operators, the necessity operator, for which I'll use the symbol $\Box$ when what is in question is *logical* necessity, cannot,

---
[1] Department of Philosophy, Victoria University of Wellington, max.cresswell@vuw.ac.nz

[2] For that reason, where my comments are about Schotch's 'reimagined' Quine, I will not refer to Quine's published works for the source of his views.

without confusion, be treated as a sentential operator (connective); since it is really a predicate like 'is valid'.

Schotch begins his discussion by taking conjunction as the paradigm, and he offers the following analysis of it:

(1)   A conjunction is true if and only if one conjunct is true and the other is true as well. (Schotch p. 32; '(1)' is my numbering.)

Schotch easily disposes of the objection that this is circular, by invoking the distinction between language and metalanguage. However, once that distinction is drawn it is clear that (1) depends on the meaning of 'and' in the metalanguage. Let's use P and Q as schematic sentences to get

(2)   'P and Q' is true (in English) iff P is true and Q is true.[3]

An instance of (2) might be the following:

(3)   'I woke up and gave my logic lecture' is true (in English) iff I woke up and gave my logic lecture.

But (3) in English is not equivalent to

(4)   'I woke up and gave my logic lecture' is true (in English) iff I gave my logic lecture and woke up.

In (4) the metalinguistic 'and' is of course the English use of 'and' to mean 'and then'. In order to make (2) correct and obvious, we need to stipulate clearly that 'and' in both the object language and the metalanguage is to be used in the truth-functional sense, to indicate that the truth or falsity of the object-language sentence is to be understood as true only in the case of the truth of both the sentences which flank the 'and'. In all other cases it is false.

Usually, however, we are engaged, not so much in giving an account of an English word like 'and' as in describing how a symbol, say $\wedge$, is to be understood in a formal language, in this case a language in which the only relevant aspect of meaning are truth values. And while we often say

(5)   '$P \wedge Q$' is true iff P is true and Q is true

we seldom say

(6)   '$P \wedge Q$' is true iff P is true $\wedge$ Q is true[4]

In (5) we are giving a stipulative definition of a symbol, and that excuses us from the question of whether or not (5) is 'correct'. We can similarly stipulate that

(7) $P \vee Q$ is true iff at least one (possibly both) of P and Q is true.

---

[3] Conditions like (2) should no doubt be stated by replacing the inverted commas by 'Quine corners', and making other adjustments to satisfy use/mention purists.

[4] The situation may be a little less clear for those who symbolise conjunction by &.

7.1. TRUTH FUNCTIONS AND NECESSITY

Rather than using sentences like (5) or (7) we can also express the meaning of ∧ or ∨ by a truth table. The standard truth tables for ∧ and ∨ are as follows:

(8)

P	∧	Q	P	∨	Q
T	T	T	T	T	T
T	F	F	T	T	F
F	F	T	F	T	T
F	F	F	F	F	F

Nothing about this is at all dubious or mysterious. Now to the case of □. Here is what Schotch says (again my numbering):

(9)   □ P is true if and only if it is necessarily the case that P is true. (Schotch, p. 34)

The question is why (9) should be any more worrisome than (2). Part of Quine's complaint is that he doesn't understand modality. This would mean that he could not accept (9) as it stands. Schotch seems to agree, and later on the same page Schotch gives us:

(10)   □ P is true if and only if P is logically true (Schotch, p. 34)

where being logically true means being valid — in the simplest case being a tautology. Schotch points out that this gives us a non-standard modal logic. Indeed it does, and we can see why by attending to Carnap's 1946 JSL paper. In what follows I have adapted slightly Carnap's notation and terminology. What I call a *PC-model*, M, is simply an assignment of a truth value, T or F, to each propositional variable. I write M ⊨ p if M assigns T to p and M ⊭ p if M assigns F to p. ⊨ can then be extended to complex formulae. The rules for truth functions are the usual ones. Thus, in the case of ∧ we have:

(11)   M ⊨ $P \wedge Q$ iff M ⊨ P and M ⊨ Q.

and so on. The rule for □ is

(12)   M ⊨ □ P iff M' ⊨ P for every PC model M'.

Given the way modal formulae are built up, the truth of □ P in any PC-model M depends only on the truth of P in PC-models. Since P is of lower modal degree than □ P, its truth or falsity in any PC-model is already defined. This answers Schotch's complaint on p. 34 that (10) cannot make sense of iterated modalities. We can say that a wff P of modal logic is valid, in this sense, iff M ⊨ P for every PC-model M. In honour of Carnap, I call this *C-validity*. It is easy to prove that every theorem of Lewis's S5 is C-valid. However, S5 is not complete for C-validity, since, where p is a propositional variable, ~ □ p is C-valid.[5] One can repair this, as Carnap appears to do, by again

---

[5] See Cresswell 2013 for these results for Carnap's modal propositional logic. C-validity can be axiomatised by adding the rule that if any non-modal wff α is not PC-valid then ~ □α is a theorem. A similar axiomatisation and completeness proof have been provided by S.K. Thomason 1973.

following some ideas of Quine. Suppose we say that a wff P of modal logic is *QC-valid* iff not only P itself but every substitution instance of P is C-valid. Then one can prove that QC-validity is exactly captured by S5.[6]

Does this then shew that Carnap has a complete answer to Quine's criticisms — at least in the propositional case? That depends on the role that modality is expected to play in deductive inference. For it could be argued that the replacement of (9) by (10) is intended to cash out an explicitly modal notion with one which does not involve modality. Thus, for instance, $P \wedge Q \models Q$ in the sense that, no matter what the truth values of P and Q are, whenever $P \wedge Q$ is true, so is P. If we translate 'and' as $\wedge$ we therefore seem to have an explanation of the intuitive validity of the inference from

(13)   Auckland is the capital of New Zealand and Christchurch is in the South Island

to

(14)   Auckland is the capital of New Zealand.

The view I want to consider is the view that truth tables give an adequate account of the intuitive validity of inferences like (13)/ (14) without reference to any semantic notion other than simple truth. Say that an operator O is truth-functional iff there is a function, f — i.e. a many-to-one relation — from truth values to truth values, such that where the truth values of $P_1,..., P_n$ are $k_1,.. k_n$ (each k_i being either T or F) then the truth value of $O P_1... P_n$ is $f(k_1,..., k_n)$. We are then assuming that the truth values here are the *actual* truth values of the sentences in question. The reason for this is that we are interested in whether or not truth tables can be used to give an analysis of logical necessity without presupposing modal notions.

The problem with this account was seen by Arthur Prior in 1964.[7] Prior introduces us to the operator 'ett', where

(15)   ' P ett Q' is true iff either P and Q or Oxford is the capital of Scotland.

Although Prior doesn't tell us, we will assume that it is a contingent matter that Oxford is not the capital of Scotland, and we will assume that Prior intended this to be so. To see the problem, compare the truth table for $\wedge$ with that for *ett*. Using (15) we may express P *ett* Q as

---

[6] Although this can be proved for propositional S5, it is, as far as I am aware, not known whether it can be proved for first-order S5. (The completeness of Carnap's predicate logic is studied in Cresswell 2014.) Unlike C-validity, QC-validity preserves the rule of uniform substitution for propositional variables. It is thus in line with the remarks Quine makes about "not taking the difference between atomic and non–atomic matrices as relevant for the logical form" on p. 349 of Creath 1990. Carnap 1946, p. 41, describes himself as 'following Quine' in using propositional variables as 'auxiliary variables' — i.e. they are not part of the object language, but are used to describe formulae of a certain sort. (Quine's views on this date from Quine 1934.)

[7] Prior 1964, p. 194. Some years later (in Cresswell 1978) I made much the same point, using a different example, in ignorance of Prior's paper. As I go on to shew in Part II, the moral Prior draws from *ett* is a little different from mine.

7.1. TRUTH FUNCTIONS AND NECESSITY

(16)   (P ∧ Q) ∨ Oxford is the capital of Scotland.

Oxford is not the capital of Scotland, which means that 'Oxford is the capital of Scotland' in (16) always receives the truth value F. (Remember that we are talking about *actual* truth values.) Thus we have

(17)
(P	∧	Q)	∨	Oxford is the capital of Scotland
T	T	T	T	F
T	F	F	F	F
F	F	T	F	F
F	F	F	F	F

This means that the table for *ett* is:

(18)
P	ett	Q
T	T	T
T	F	F
F	F	T
F	F	F

In other words it is *exactly the same* as the table for ∧, *for every argument*. So if logical validity means truth-preservation no matter what (actual) truth values *P* and *Q* have, we have that the passage from *P ett Q* to *P* is a logically valid schema; and if we take an inference to be valid if it has the form of a valid schema we have that any inference of that form is logically valid. In particular the inference from

(19)   Auckland is the capital of New Zealand ett Christchurch is in the South Island

to

(20)   Auckland is the capital of New Zealand

is valid. But, as we are assuming, it is contingent that Oxford is not the capital of Scotland. Suppose it were the capital. Then by (15), (19) would be true. But (20) would still be false, and so the inference from (19) to (20) is intuitively invalid, in the sense that it is possible (even if it is not actually so) to have a true premise and a false conclusion.

It might be argued that ∧ is a logical constant while *ett* is not, and indeed this may be so. But what is the difference? The difference is surely that *if Oxford had been the capital of Scotland* then the truth table for *ett* would have been:

(21)
P	ett	Q
T	T	T
T	T	F
F	T	T
F	T	F

The difference between the truth tables for ∧ and for *ett* is that although they coincide in the actual world, they differ in worlds in which Oxford is the capital of Scotland.[8] Unlike the table for *ett* the table for ∧ remains the same in *every* world, and *that* is what makes ∧ a logical constant. Further, that is what makes the inference from (13) to (14) *necessary*. One can of course argue that truth-functionality should be independent of what the actual world is like, so that Prior's argument may not really be an argument against truth-functionality as such, and in a sense that is true.[9] But that reading is only legitimate if one is interpreting the tables as remaining constant whatever the world might happen to be like. And if one does that then one is presupposing a modal component. Undoubtedly, if you are not worried by the fact that the necessity of truth-functionally valid sentences involves a modal notion, then Prior's argument is irrelevant. But it is relevant to Quine's criticism of Carnap, because Carnap was trying to claim that you can obtain logical necessity, for at least some sentences, simply from the notion of truth alone. Carnap's semantics will not help us to distinguish ∧ from *ett*.

7.2 Prior on Truth

The paper in which Prior's 'ett' example occurs (Prior, 1964) is concerned to take a much more radical position than the one I have urged in my discussion of Schotch's views, and this part of the paper examines Prior's views on the connection between truth and semantics. If my exposition is persuasive, what Prior offers is an objection to a whole way of thinking about logic which is presupposed in the discussion in Part I, since it calls into question the role of truth as it has been applied in current work.[10] I will begin by looking at Prior's critical notice in *Mind* of Tarski 1956.

> When the presupposed definition of 'satisfaction' is examined, however, it will be found that this definition of truth has a further defect. Satisfaction can only be defined in the following roundabout way (I again give the thing roughly): ' x is included in y' is satisfied by the pair of classes a, b if and only if a is included in b; 'not– y' is satisfied by any group of classes which does not satisfy y; ' x or y' is satisfied by any group of classes which either satisfies x or satisfies y; and a function preceded

---

[8]Schnieder 2008, p. 67, discusses a sense in which an operator could be said to be truth-functional provided it is always defined by a truth table, though not always by the *same* truth table. *ett* is truth-functional in this weaker sense.

[9]This is, in effect, the position advocated by Hill and McLeod 2010 in reply to Schnieder 2008, who produces an argument similar to Prior's. Prior's argument, however, has the advantage of not involving 'that'-clauses. It is perhaps of some interest to take a look at how introductory logic texts define truth-functionality. Probably the commonest is to say that the truth of the complex is 'determined by' or 'depends on' nothing more than the truth values of its components. In an examination of 16 arbitrarily selected introductory texts, Rebecca James, (a student of Adriane Rini's) found five which used the notion of a function, while the remainder used notions like 'depends on', 'is determined by' or refer to what is 'known'. Only two explicitly acknowledged that these are modal notions, in that they assume that the truth tables will remain constant whatever the world is like. One of the introductions which used the notion of a function said 'is a function of (is completely determined by)' as if these were equivalent.

[10]In order to avoid misunderstanding I should perhaps stress that I am not claiming to *endorse* Prior's attitude to truth-conditional semantics. In fact I think it is misguided. But it seems to me that it *is* his attitude.

7.2. PRIOR ON TRUTH

> by the universal quantifier is satisfied, etc. Such a piecemeal definition of satisfaction means a similarly piecemeal definition of truth, when it is all spelt out; and the more complex the language considered the more pieces there will be. I know there are plenty of quite un-Tarski–like people who will be entirely happy about this — people who contend that even in 'everyday or colloquial' language the word 'true' has different meanings when applied to sentences of different sorts: so that it can have a single meaning only in the sense of a disjunction of these. My own understanding of ordinary language is quite otherwise; there are no doubt dozens of different ways of deciding whether a given sentence is true, but what it means for a sentence to be true is pretty much the same throughout, and pretty much what was suggested at the beginning of this discussion. (Prior 1957, p. 408f.)

What Prior had suggested was

> We generally say that a sentence is true if it says that something is so and it is so. (Cf. Prior 1957, p. 406)

Prior holds that this needs refinement because of the semantic paradoxes, but contrasts it with the 'piecemeal definition of truth' which he attributes to Tarski. What he seems to have in mind in mind is this. Tarski appears to have thought that he was defining *truth*. One feature of Prior's view of Tarski is that Tarski's recursive specification of the truth conditions of sentences of a formal language in fact depends on already knowing what it means to say that a sentence is true or is false; since otherwise it's not clear what Tarski's formal definition is a definition *of*. In its way it is the same criticism as that which Quine brought against Carnap's definition of analyticity in a particular language.[11] In Carnap 1963 Carnap replies to Quine's attack on analyticity. This reply occurs just after the section (pp. 889–899) where Carnap has been expounding his latest conception of the logic of modalities. While Quine can accept that Carnap may have defined a formal concept which he calls 'analyticity in L', where L is a formal language, Quine cannot accept that there is any intuitive natural language concept which can be the explicandum. Carnap sums it up like this:

> My interpretation of Quine's intention is as follows, formulated in my terminology. It seems to me that Quine's criticism is not directed against the proposed semantical explicata. I believe that he would agree that, e.g., my rules of the above mentioned kinds, leading to the definition of "A–true", are in themselves exact and unobjectionable. His criticism is rather that there is no clear explicandum, in other words, that the customary pre–systematic explanations of analyticity are too vague and ambiguous, and basically incomprehensible. This would make it understandable why he requires for analyticity an empirical criterion, while he does not require it for truth. In Quine's view, there is the following basic difference. In the

---

[11] For Quine's objections see Quine 1953, and for Carnap's reply see Carnap 1963, pp. 919-921.

case of truth he recognizes a sufficiently clear explicandum; i.e., before an explication had been given, the use of this concept had been sufficiently clear, at least for practical purposes. (Carnap 1963, p. 918f.)

What Carnap is telling us is that although his formal account of analyticity in L may well be a completely adequate and precise definition of what it is for a formula to be 'analytic in L', it will not help him answer Quine unless he is able to tell us what concept of ordinary language it is an explication of. Quine had pointed out that nothing in Carnap's definition has anything to offer about what it is that Carnap is supposed to be defining. Another way of putting Quine's point is that Carnap cannot be regarded as *defining* analyticity since we must *already* have such a notion if 'analytic' in 'analytic in L' is to mean the same for every L. Carnap's reply was that, yes, Quine is correct in that his account depends on there being a phenomenon which we all recognise, but in fact there *is* such an empirical phenomenon. In the case of truth, at least according to Carnap, Quine does seem to recognise that we just do have a notion of what it is to say that a sentence is true or false, and if we do then presumably we would be entitled to use truth in our study of language. By contrast, what Prior's point in Prior 1957 amounts to is that the same comments as Quine made about analyticity apply to the notion of truth itself.[12] So suppose our question is: What does it *mean* to say that a sentence S is true or is false? Well, one might say something like this, that it means that things are as S says they are. That is to say, for S to be true, two things must hold:

(22) S says that p

and

(23) p

One might express (23) as

(24)   It is true that p

though Prior 1971, p. 11, following F.P. Ramsey,[13] seems clear that he regards (24) as no different in meaning from (23), which is why it is called the 'no truth' theory. But whether or not (23) and (24) really are different in meaning, it doesn't help with what to say about (22).

Much current work in semantics, either implicitly or explicitly, trades on the idea that the job of a semantic study of the languages of formal logic is concerned with truth conditions. Wittgenstein's *Tractatus Logic–Philosophicus* (Wittgenstein 1922) contains the following statement:

4.024 To understand a proposition means to know what is the case, if it is true. (One can therefore understand it without knowing whether it is true or not.) One understands it if one understands its constituent parts.

---

[12] Similar points are made against Tarski in Field 1972.
[13] See Ramsey 1927 (especially page 39 of the reprint in Ramsey 1990.)

7.2. PRIOR ON TRUTH

One can take this, and the surrounding passages, as Carnap for one did (Carnap 1947, p. 9f., also Davidson 1967), as endorsing the view that meaning can be understood in terms of truth conditions. Prior discusses this view in Prior 1979, a work written in the early 1950s:

> Some modern logicians, following Wittgenstein, argue that two sentences have the same meaning, and are consequently instances of the same proposition, if they have the same 'truth–conditions', i.e. if any possible circumstances which would make either of them true or false would have the same effect on the other. On this view, we would need to say, for example, that 'It is raining' is the same proposition as 'Either it is raining but not snowing or it is doing both', for any possible state of affairs in which the former is true is one in which the latter is, and vice versa, and similarly with falsity. (If it is doing both, both the simple and the compound proposition are true; if it is raining but not snowing, both are true; if it is snowing but not raining, both are false; and if it is doing neither, both are false.) Objections to this are (a) that it seems to involve 'propositions in themselves' (as well as the propositions that are sentences), under the guise of 'possible states of affairs'; and (b) that the suggested usage would make it always false to say that two different propositions logically imply one another, since by the criterion suggested, this relation would automatically make any such allegedly different propositions the same proposition. (For if each implies the other, any circumstance under which either would be true, would be one under which the other would be.) (Prior 1979, p. 36f.)

The difficulty that Prior alludes to here has long been known as a problem for the truth-conditional theory of meaning. Luckily though, for looking at Prior's problem all that is necessary is to acknowledge that even if the meaning of S is *more* than the conditions under which S is true, yet surely it must *at least* be so that the meaning of S (assuming that S is the kind of sentence for which it makes sense to speak of its truth and falsity) must somehow determine the conditions under which it is true. Take a sentence like:

(25)   An apple is on the sideboard.

You don't need to know English to know what it is for an apple to be on the sideboard. But you *do* need to know English to know that what (25) says is true provided that an apple is on the sideboard, and is false otherwise. Just what this knowledge consists in is a difficult question (considered at length in Cresswell 1994), but at least it is what underlies the ability for a person who wants to get an apple to know where to look. A similar account underlies the sentences

(26)   An apple is on the sideboard and an orange is on the sideboard.

and

(27)   An apple is on the sideboard or an orange is on the sideboard.

One can imagine either (26) or (27) being offered in answer to the question

What kind of fruit can I take to school?

Suppose that there is a language English*, which is just like English except that 'and' in English* means what 'or' does in English and 'or' in English* means what 'and' does in English. Whatever theory of meaning you hold, it seems obvious that the difference between English and English* is that the way an English speaker expects the sideboard to look if (26) is true is just the way an English* speaker expects it to look if (27) is true, and the way an English speaker expects the sideboard to look if (27) is true is just the way an English* speaker expects it to look if (26) is true.[14]

Imagine a semanticist among the speakers of English*, Alfreda Tarski*. Tarski* is very impressed by the work of semanticists of English, and noting the way semanticists of English treat 'and' and 'or' decides to do the same for these words in English*. She also notes that some semanticists of English prefer to replace 'and' by $\wedge$ and 'or' by $\vee$, and she decides to do the same. Further, Tarski* uses T and F and provides exactly the same 'truth' tables as those provided by the semanticists among the English speakers for these words — i.e., the ones given in (8) — and she is hailed as successfully defining truth. However, some English* speakers, among them Prior*, object that while Tarski* may have succeeded in defining something, it cannot be truth, since it does not respect the connections which hold between speakers of English* and situations about apples and oranges on sideboards. Tarski*'s reply is that, because natural language is inconsistent, truth is being defined for a formal language. Prior* appears to agree, and comments:

> For everyday or colloquial language just isn't the sort of language in which the sense of its sentences is entirely determined by the sense of their parts and the ways in which these parts are put together. It is a definition of truth which can be applied to this last sort of language that Tarski* is seeking; when that is understood, and only when that is understood, her achievement can be seen in its right proportions. (See Prior 1957, p. 408, *mutatis mutandis*.)

Now to Prior 1964. Its title refers to the earlier Prior 1960, which discusses the view that the meaning of a logical operator is to be specified in terms of inference rules in the form of introduction and elimination rules. But the 1964 paper makes a link with truth tables. The first paragraph of Prior 1964 reads:

> 1. It is one thing to define 'conjunction–forming sign', and quite another to define 'and'. We may say, for example, that a conjunction-forming sign is any sign which, when placed between any pair of sentences P and Q forms a sentence which may be inferred from P and Q together, and from which

---

[14] Some might be worried by the fact that I have described the situation in which these speakers find themselves when looking at the sideboard, by using words like 'and', but I have already pointed out that, as Schotch has noted, an adherence to the language/metalanguage distinction explains why this worry is misplaced. What kind of fruit is on the sideboard is quite independent of the meaning of 'and' or 'or'.

7.2. PRIOR ON TRUTH

we may infer P and infer Q. Or we may say that it is a sign which, when placed between any pair of sentences P and Q, forms a sentence which is true when both P and Q are true, and otherwise false. Each of these tells us something that could be meant by saying that 'and' for instance, or '&' is a conjunction–forming sign. But neither of them tells us what is meant by 'and' or by '&' itself.

I have nothing to say about definitions by means of inference rules, but I *am* concerned with what Prior says about truth conditions, since this connects with what was said about truth tables like those set out in (8) above. The second paragraph of Prior's article reads:

2. Moreover, each of the above definitions implies that the sentence formed by placing a conjunction–forming sign between two other sentences already has a meaning. For only what already has a meaning can be true or false (according as what it means is or is not the case), and only what already has a meaning can be inferred from anything, or have anything inferred from it. (Prior 1964, p. 159)

In order to understand what Prior is up to we need to ask what he has in mind by his phrase 'already has a meaning'. In Section 1 his complaint is that truth tables cannot tell us "what is meant by 'and' or by '&' itself", and this suggests that Prior is claiming that, in some mysterious way, advocates of the truth-table account of the meaning of 'and' think that a sentence like (26) does not have a truth value until the truth table has supplied it. But of course this is a silly picture of what the truth table account of the meaning of 'and' amounts to. For the truth table account of the meaning of 'and' only makes sense in the first place when applied to sentences with truth values. This is why I began Part II with Prior's criticism of Tarski's theory when understood as a theory of what *constitutes truth*. For then Prior's complaint is that it cannot be so that *truth* means something different when applied to (26) from what it means when applied to (27). For if 'true' does mean something different in each case it is not clear how the difference between 'and' in English and 'and' in English* can be illustrated by the fact that English is captured by the truth table for ∧ in (8) above, while 'and' in English* is captured by the table for ∨. In order for these tables to explain the difference between English and English* we need to be assured that T and F refer to the same *truth* value, whether we are talking about English or about English*. English and English* speakers do not disagree about the meaning of 'true'. They disagree about the meaning of 'and' and 'or'.

Why did Prior not see this? Prior 1957 makes it clear that he did not accept that Tarski had succeeded in defining *truth*. But why did he assume that therefore truth-tables could not be used in defining the meaning of words like 'and' or 'or', given that we *have* a notion of truth? My guess is that he never took seriously the connection between truth and meaning. Because he dismissed the truth-conditional theory of meaning — since he felt that sameness of truth conditions is not *sufficient* for sameness of meaning — he failed to see the importance of the fact that sameness of truth conditions might be *necessary* for sameness of meaning. This seems indicated in his remarks comparing 'inferential definitions' and 'truth tables', where on p. 160f. he says

> Unlike Stevenson, l can see no difference in principle between these devices.(Prior 1964, p. 161f)[15]

The comment made above about Section 2 can also be made about Section 7 which reads:

> 7. But these indirect and informal ways of fixing the sense of a word, however indispensable they may sometimes be, have definite limitations. In the first place, a definition of a class of signs, as opposed to the definition of a particular sign itself, may mark off a class which turns out to be empty. There are in fact *no* contonktion–forming signs in either of the senses given to this phrase in Section 4; and the information that 'tonk', or anything else, is such a sign, is simply false. 'Contonktion–forming sign', like 'present King of France', is a perfectly clear description which applies to nothing whatever. (*op cit*, p. 161)

A 'contonktion sign' is defined by truth tables. Specifically it is

> one which, when placed between any two sentences P and Q, forms a sentence which is true if P is true, and false if Q is false (and therefore of course both true and false if P is true and Q is false.) (*op cit*, p. 160.)

and it is clear that (if, presumably like Prior, we are assuming a classical notion of truth in which no proposition can be both true and false) there is no truth function which satisfies the conditions for contonktion, so that, on the assumption that we already have the notion of truth, and are defining the meaning of an operator in terms of it, *this* definition does not characterise anything. What this does appear to shew is that *provided that we already acknowledge that sentences have truth values* — where this doesn't just mean that we have stipulated two 'values' but that we are talking of genuine truth — the impossibility of a condition like that for contonktion, does not cast doubt on the claim that truth tables can define the meaning of truth–functional operators. So why should Prior think of it as a *limitation* of the truth-table method that there is no such operator? Presumably his worry is because he thinks that truth-tables are supposed to define *truth* itself, a notion we do not already have. And his argument is that if we do not already have the notion of truth then we cannot use facts about inconsistency to rule out truth tables for operators like that for contonktion. Otherwise, as Prior points out right at the beginning of the article, we are involved in a 'purely symbolic game', and it is the burden of Prior 1964 to remind us that if that is so then it has no connection with truth or meaning.

---

[15]The reference is to Stevenson 1961, and the sentence Prior has in mind is probably:
The important difference between the theory of analytic validity as it should be stated and as Prior stated it lies in the fact that he gives the meanings of connectives in terms of permissive rules, whereas they should be stated in terms of truth–function statements in a meta–language. (*op cit* p. 127)

7.3 Conclusion

In this paper I have looked at two criticisms of the truth-conditional theory of meaning advocated by Wittgenstein and Carnap. One criticism from the left and one from the right. Part I examines Schotch's discussion of Quine's criticism of propositional modal logic, and argues that even the apparently simple semantics for truth-functional operators has more intensionality in it than at first might appear. An essential component in this enterprise is the role of Prior's *ett* operator. But the use Prior puts this operator to appears to be an attack on the whole idea of looking at meaning in terms of truth conditions. Part II examines the arguments in Prior 1964 in terms of the role of truth conditions in semantics, and suggests that Prior's arguments can only be seen to make sense as a criticism of the view that work like that of Tarski tells us something about what *constitutes* truth.

Schotch has pointed out some challenges which a reimagined Quine might present for modal logic. It is I think undeniable that Quine was right that you cannot obtain an account of modal notions like logical *necessity* and logical *possibility* using truth as your only semantic notion. Quine's position was to eschew modality. The price to be paid in that case is of course that you cannot then explain how, in a deductively valid inference, if the premises are true the conclusion *must* be true, and that this needs no factual import. So, the evaluation of Quine's views on modal logic appears in the end to turn on whether you think that a valid deductive inference requires such a guarantee. In the case of Prior, the question turns on whether you think, as Prior seems to have thought, that a truth-conditional theory of meaning is supposed to be also a theory of what truth itself is in the first place, rather than a theory of how an already understood notion of truth and truth conditions can be used to specify the meaning of such words as the connectives of propositional logic.

Appendix: The Operator 'ett$_3$'

One objection that may be raised to the use I made of the 'ett' operator in Part I is that there is the following asymmetry between 'and' and 'ett' in that, while you can define the meaning of 'ett' in terms of 'and' and 'or' — and indeed this may be the only way for us to understand its meaning — you cannot proceed in the reverse direction and define 'and' or 'or' in terms of 'ett'. That is indeed so, since if Oxford *were* the capital of Scotland, p ett q would be true no matter what propositions p and q are, and in such a case 'ett' would not suffice to define any non-trivial truth function.

The first reply is that this does not affect the use to which I have put the discussion of 'ett'. For the purport of that discussion is not to claim that there is no difference in logical status between 'and' and 'ett'. Quite the reverse in fact. For the meaning of 'and' can be defined by a truth table which remains constant however the facts may turn out to be, while the table for 'ett' is contingent. So it is only to be expected that there may be significant logical differences between sentences with 'and' and 'sentences with 'ett'. The point is that these differences do not shew up in a truth table which only considers actual truth values. They only shew up when we imagine that things might be otherwise.

Nevertheless, there is an interesting formal question here. Even if *ett* cannot be used to define truth functions, can you find an operator which is like ett in having a truth table which differs from world to world, but which, despite this, *is* adequate to defining the truth functions? In fact the answer is yes. Consider an operator I shall call 'ett$_3$'.

(28)   For any three propositions p, q and r, ett$_3(p, q, r)$ is true iff either Oxford is the capital of Scotland and p, q and r are all false, or Oxford is not the capital of Scotland and p and q are not both true, and r is false.

If we take the standard truth functions as given (28) can be expressed as an equivalence in the following way. Use O to mean 'Oxford is the capital of Scotland', and assume the truth functors $\sim$, $\wedge$ and $\vee$. Then

(29)   ett$_3(\alpha, \beta, \gamma) \equiv (O \wedge (\sim \alpha \wedge \sim \beta \wedge \sim \gamma)) \vee (\sim O \wedge \sim (\alpha \wedge \beta) \wedge \sim \gamma)$

In the real world Oxford is not the capital of Scotland, and so, in the real world the truth table for ett$_3$ is

ett$_3$	(α,	β,	γ)
F	T	T	T
F	T	T	F
F	T	F	T
T	T	F	F
F	F	T	T
T	F	T	F
F	F	F	T
T	F	F	F

which is the same as that for

$\sim O$	$\wedge$	$\sim (\alpha \wedge \beta)$				$\wedge$	$\sim \gamma$	
TF	F	F	T	T	T	F	F	T
TF	F	F	T	T	T	F	T	F
TF	F	T	T	F	F	F	F	T
TF	T	T	T	F	F	T	T	F
TF	F	T	F	F	T	F	F	T
TF	T	T	F	F	T	F	T	F
TF	F	T	F	F	F	F	F	T
TF	T	T	F	F	F	T	T	F

no matter what α, β and γ are. So we have that

   $(\sim(p \wedge q) \wedge \sim r) \supset$ ett$_3(p, q, r)$

is PC-valid, in the sense that it has the value T no matter what (actual) truth values are given to p, q and r. As a consequence, if we use $\alpha, \beta \vDash \gamma$ to mean that no combination of truth values to the variables of α, β or γ ever makes it so that α and β are both true while γ is false, then

7.3. CONCLUSION

(30) $\sim(p \wedge q), \sim r \models \text{ett}_3(p, q, r)$

But suppose that Oxford *were* capital of Scotland. Then, in such a case,

(31)   Wellington and Edinburgh are not both of them the capital of New Zealand, and nor is Paris

could still be true; and

   None of Wellington, Edinburgh or Paris is the capital of New Zealand

could still be false. That is to say, there is a world in which (31) is true and

   Ett$_3$(Wellington is the capital of New Zealand, Edinburgh is the capital of New Zealand, Paris is the capital of New Zealand)

is false. This means that (30) does not respect intuitive validity.

The next task is to demonstrate the expressive power of ett$_3$. For this we make use of the dagger operator ↑, where ↑ is defined by the truth table:

α	↑	β
T	F	T
T	F	F
F	F	T
F	T	F

↑ (like the Sheffer stroke |) is sufficient to express all truth functions. And what is significant for this appendix is that, if we take ett$_3$ as primitive, ↑ may be defined as:

(32) $\alpha \uparrow \beta =_{df} \text{ett}_3(\alpha, \alpha, \beta)$

We can see that (32) is a good definition by comparing it with the reverse definition stated in (29) based on primitive truth functors. When you apply (29) to ett$_3(\alpha, \alpha, \beta)$ you get

$$(O \wedge (\sim \alpha \wedge \sim \alpha \wedge \sim \gamma)) \vee (\sim O \wedge \sim(\alpha \wedge \alpha) \wedge \sim \beta)$$

which is easily seen to be equivalent to $\alpha \uparrow \beta$ in the strong sense that the equivalence holds no matter what world we are in. This means that all truth functions can be expressed by taking ett$_3$ as primitive. Not only is ett$_3$ sufficiently powerful to define, on its own, all the truth functors. It is also sufficiently powerful to define O. For assume that $\bot$ is a symbol which is always (necessarily) false — which may be defined as $p\uparrow(p\uparrow p)$; and $\top$ — which may be defined as $(p\uparrow(p\uparrow p))\uparrow(p\uparrow(p\uparrow p))$, and is always true. Then

$O \equiv \sim\text{ett}_3(\top, \bot, \bot)$

is not only true but is necessarily true. For it is equivalent to

$$O \equiv \sim((O \wedge \sim \top \wedge \sim \bot \wedge \sim \bot) \vee (\sim O \wedge \sim(\top \wedge \bot) \wedge \sim \bot)).$$

So the ett$_3$ operator is like the ett operator in the sense that it licenses tautologies made from truth tables which respect actual truth values, yet which do not represent intuitively valid inferences. But in addition, unlike ett, ett$_3$ is sufficiently powerful *on its own* to define all truth functions. By this latter is meant that the definition of the truth functions in terms of ett$_3$ holds *independently of which world we are in.*

7.4 Bibliography

Carnap, R. (1946). Modalities and quantification. *Journal of Symbolic Logic 11*(2), 33–64.

Carnap, R. (1947). *Meaning and Necessity*. Chicago, University of Chicago Press.

Carnap, R. and P. A. Schilpp (1971). Intellectual autobiography. *Journal of Symbolic Logic 36*(1), 178–179.

Creath, R. (1990). *Dear Carnap, Dear Van: The Quine-Carnap Correspondence and Related Work: Edited and with an Introduction by Richard Creath*. Berkeley, University of California Press.

Cresswell, M. J. (1978). Semantics and logic. *Theoretical Linguistics 5*, 19–31. Reprinted in Semantical Essays: Possible Worlds and Their Rivals, Kluwer Academic Publishers, 1988, pp. 34-46.

Cresswell, M. J. (1994). *Language in the World: A Philosophical Enquiry*. Cambridge, Cambridge University Press.

Cresswell, M. J. (2013). Carnap and McKinsey: Topics in the pre-history of possible worlds semantics. In J. Brendle, R. Downey, R. Goldblatt, and B. Kim (Eds.), *Proceedings of the 12th Asian Logic Conference*, pp. 53–75. World Scientific.

Creswell, M. J. (2014). The completeness of carnap's predicate logic. *Australasian Journal of Logic 11*, 47–63.

Davidson, D. (1967). Truth and meaning. *Synthese 17*(1), 304–323.

Hill, D. J. and S. K. McLeod (2010). On truth-functionality. *Review of Symbolic Logic 3*(4), 628–632.

Prior, A. N. (1957). Critical Notice of Alfred Tarski *Logic, Semantics and Metamathematics. Mind 66*, 401–410.

Prior, A. N. (1964). Conjunction and contonktion revisited. *Analysis 24*(6), 191–195. [Reprinted in (Prior, 1976b). Page references are to the reprint.]

Prior, A. (1967). The runabout inference ticket. *Analysis*, pp. 38–9. [Reprinted in (Prior, 1976b). Page references are to the reprint.]

Prior, A. N. (1971). *Objects of Thought*. Oxford, Clarendon Press.

Prior, A. N. (1976a). *The Doctrine of Propositions and Terms*. London, Duckworth.

Prior, A. N. (1976b). *Papers in Logic and Ethics*. London, Duckworth.

Quine, W. V. (1934). Ontological remarks on the propositional calculus. *Mind 43*(172), 472–476.

Quine, W. V. O. (1951). Two dogmas of empiricism. *Philosophical Review 60*(1), 20–43. [Page references are to the reprint in (Quine, 1953).]

Quine, W. V. (1953). *From a Logical Point of View*. Harvard University Press.

Ramsey, F. P. (1927). Facts and propositions. *Proceedings of the Aristotelian Society 7*(1), 153–170. [Page references are to the reprint in (Ramsey, 1990).]

Ramsey, F. P. (1990). *F.P. Ramsey: Philosophical Papers*. Cambridge University Press.

Schnieder, B. (2008). Truth-functionality. *Review of Symbolic Logic 1*(1), 64–72.

Schotch, P. K. (Forthcoming) *Essays in Philosophical Logic*

Stevenson, J. T. (1961). Roundabout the runabout inference-ticket. *Analysis 21*(6), 124–128.

Tarski, A. (1936). The concept of truth in formalized languages. In A. Tarski, *Logic, Semantics, Metamathematics*, pp. 152–278. 1956 Oxford University Press.

Thomason, S. K. (1973). A new representation of $S5$. *Notre Dame Journal of Formal Logic 14*(2), 281–284.

Wittgenstein, L. (1922). *Tractatus Logico-Philosophicus*. London, Kegan Paul.

Chapter 8
Globalization Makes Inconsistency Unrecognizable

John Woods[1]

8.1 Ex Falso Quodlibet

8.1.1 Contradictions: True and Provable

Trivialism, near-trivialism, dialethism and consistentism are doctrines about contradictions. Each has a so-called semantic version and a so-called syntactic one. I'll explain "so-called" in due course.

Semantically, trivialism asserts that everything is both true and false; near-trivialism asserts that nearly everything is both true and false; dialethism asserts that almost nothing is both true and false; and consistentism asserts that nothing whatever is both true and false. (I mean, of course, concurrently and unambiguously true and false in all the same respects.) Syntactically, trivialism, near-trivialism, dialethism and consistentism make the same assertions, not about truth and falsity, but rather about technical senses of provability and unprovability.

It is necessary that we not make too little of the fact that the doctrines currently in view come in these two versions. For some time now logicians have conformed their thinking to a certain quite common idea. It says that, in the more successful of our mathematically well-built logistic systems, the semantic and the syntactic go hand in hand, by way of soundness and completeness metatheorems. Actually, whether they do or not depends on how we understand the provability predicate. If it is taken in its intuitive sense that is, in the sense in which it has been invoked by the working mathematician from Thales onwards and consistentism fails for the true and false it likewise fails for the provable and unprovable, i.e. intuitively. For in the sense in which it is taken by the working mathematician, something is provable only when it is possible

---

[1]The Abductive Systems Group, Department of Philosophy, University of British Columbia, www.johnwoods.ca; john.woods@ubc.ca.

to demonstrate conclusively that it is true. Whereupon, if nothing whatever is true as opposed to false, nothing whatever is both provable as opposed to not. Syntactic conceptions of proof are another matter. Syntactic proofs bear no intrinsic tie to the working mathematician's intuitive notion. The proofs of proof theory aren't demonstrations of intuitive truths. They are syntactic regulators for the placement of their own terminal lines. When we say that a string of a formal language is provable in a system, we mean something like "is generable under the system's syntactic transformation rules". Although there is no intrinsic tie to the truths sought by the working mathematician, sometimes the strings in the extension of a system's proof-predicate stand in a close correspondence to those advanced by its model theory as truths of logic. But it is by no means a given that system's truths of logic will stand one-to-one with a corresponding subset of the mathematical truths of the mathematician's working language English, German or whatever else. [2] However, for the business at hand, it suffices to emphasize that when provability is taken in its purely syntactic sense, it is certainly not the case that syntactic consistentism is true even when semantic consistentism is. All that this comes down to is that there are some such systems in which a given formula and its syntactic negation are both generable under the terminal-line placement rules. When this happens, the systems are negation-inconsistent. The question is, are they also absolutely inconsistent?

8.1.2 The Lewis-Langford Proof

There is a problem with consistentism, which is also a problem for dialethism and near-trivialism. There exists a good-looking proof that if anything is both provable and not, that is, negation-inconsistent, then everything is both provable and not, that is, absolutely consistent. The theorem that this proof proves is widely known as ex falso *quodlibet*, which is believed to have originated as early as 1200 with Alexander Nekham of the School of Cologne, which actually was based in Paris. [3] A prominent modern version was advanced by Lewis and Langford in 1932.[4] The proof is easily seen to be valid in classical logic and all the modal logics on whose behalf the modern proof was advanced in the first place. [5] It is a very simple proof, effortlessly set out in the elementary truth functional logic of propositions:

---

[2] Even if it did it leaves undealt with the question of how a predicate of an uninterpreted sequence of marks, conveying that the sequence in question has a model in all set-theoretic interpretations gets to represent a predicate of a natural sentence, such that whenever "'S' is true" holds, so does S. I'll have more to say of this below.

[3] Alexander Nekham, *De Naturis Rerum*, T. Wright, editor, London: Longman, 1863, pp. 288-289. I owe this reference to Stephen Read, *Relevant Logic: A Philosophical Examination of Inference*, Oxford: Blackwell, 1988, p. 31, n. 10, 11.

[4] C. I. Lewis and C .H. Langford, *Symbolic Logic*, New York: Appleton-Century Croft, 1932. Reprinted in 1959 by Dover; p. 259.

[5] In the Lewis systems S1-S5, consequence is defined as follows: B is a consequence of A iff $\sim \diamond(A \wedge \sim B)$, a condition that is satisfied for any impossible A (and also for any necessary B).

8.1. EX FALSO QUODLIBET

1. $A \wedge \sim A$ assumption
2. A 1, $\wedge$-elim
3. $A \vee X$ 2, $\vee$-intro for arbitrary X
4. $\sim A$ 1, $\wedge$-elim
5. X 3, 4, DS

If semantic consistentism were true, the other three would be false. If semantic consistentism were false, then at least one statement would be true and false. If that were so, trivialism would be sanctioned by ex falso, and near-trivialism and dialethism would fail. The received wisdom is that consistentism is not false, that no statement whatever is both true and false. If that is so, there is little to fear from ex falso. Syntactic consistentism is a different story. Axioms are theorems and derivability is a theorem-preserving relation. When the axioms and proof rules of a system sometimes generate both A and its negation $\ulcorner \sim A \urcorner$, the resultant negation-inconsistency is met with alarm. If *ex falso* is true, alarm spreads to panic. For now everything is provable there. Syntactic consistentism is certainly false. If ex falso is true, so is syntactic trivialism true and syntactic near-trivialism and dialethism false.

Some people think that the Lewis-Langford proof is but a quirk of the extensionality of the connectives that underpin its line-to line progression. Their point seems to be that quirky connectives are more or less bound to patronize some unpleasant theorems. Accordingly, the bad news that the *ex falso* proof is thought to deliver reflects badly not on how the consequence relation actually works, but on a poor choice of connectives can go so seriously awry. The proof's authors saw this objection coming. They said the proof captured the intuitive meanings of inference and proof, and challenged their critics to pick any of the proof's rules whose validity clashed with those intuitive meanings. Let me see if I can cash this out a bit.

Consider the following argument, which I set out in plain English without reliance on truth functional connectives or the word "or", for arbitrary English statement-expressing sentences $S, \ldots, X$.

(1) S and not-S by assumption

Then on the principle that from the assumption of both each follows, it follows from (1) that

(2) S.

On the principle that if some sentence follows so does at least one member of any set containing it and any other sentence,

(3) At least one of $\{S, X\}$ follows, for arbitrary X.

By the reasoning that took us from (1) to (2) we now arrive at

(4) Not-S.

Then on the principle that if at least one of two statements follows, which is what (3) says, and it's not this one, which is what (4) says, then it's the other that follows, we obtain

(5) X

via the negation rule.[6]

It is interesting to focus on what Lewis and Langford were not claiming to have proved. They were not claiming merely that ex falso was syntactically generable in classical logic and modal systems such as S1-S5. They were claiming that their proof proved that ex falso is true, period; that it is an objective fact about deductive consequence as it actually is in its own right, that it really does deliver ex falso. This may strike the modern pluralistic ear as quaintly old-fashioned. But Lewis and Langford meant what they said, acknowledging in effect the convergence of truth on being.

Provability is one of those properties much prized by both the theorist and the everyday reasoner on the ground. So, too, is consistency. Sometimes consistency is lost unawares. If ex falso is true, even the innocent loss of consistency does no damage to provability; I mean, of course, syntactic provability. Provability is a good thing to have. But if ex falso is true, provability is radically too much of a good thing, hence a radically bad thing. When this sort of detonation[7], occurs I will say that negation-inconsistency visits upon provability the *globalization effect*. When globalization strikes, properties that we value go viral. So do properties we don't value, e.g. negation-inconsistency itself. True belief is another of the good things we prize. Sometimes we adopt a belief from something we already believe because the new one follows from the old one. If, unbeknownst, our belief-system BEL is negation-inconsistent, there will be cases in which its following from an A we believe is reason for believing some new belief B. In every such case the negation $\ulcorner \sim B \urcorner$ of the new one also follows from something (else) in BEL. At least so it does if belief is closed under consequence and we believe—albeit implicitly—the totality of what we believe. For in that case we would have it that BEL is inconsistent, that BEL proves $\ulcorner \sim B \urcorner$ and that you believe all the propositions expressible in BEL. It is widely accepted by formal epistemologists and others that these are legitimate conditions on rational belief-change. I happen not to be in their number, for reasons set out in my *Errors of Reasoning: Naturalizing the Logic of Inference*.[8]

---

[6] It might be objected that since S follows and not-S also does, it can't be said that in its occurrence at (4) not-S lacks the power to expel S in its occurrence at (3). This is a move as old as the Middle Ages, embodying the idea that if both S and not-S are true, neither can contradict (or eliminate) the other. Whatever its merits, this is not an available option here. It might be that the joint truth of S and not-S subdues the eliminative force of the negation rule, but not surely their joint consequence from a contralogical assumption. If we performed a "true"-"false" valuation scan on S any X, then when (3) and (4) come out true, X is true. When (3) comes out false. S is false and X is false (for any X).Negation rules at line (5). See further section 5 below.

[7] Peter Schotch and Ray Jennings, "On detonating", in Graham Priest, Richard Routley and Jean Norman, editors, Paraconsistent Logic, pages 302-327, Munich: Philosophia Verlag, 1989. For some important improvements see Dorion Nicholson and Bryson Brown, "Representation of forcing", in Peter Schotch, Bryson Brown and Raymond Jennings, editors, *On Preserving: Essays on Preservationism and Paraconsistent Logic*, pages 145-160, Toronto: University of Toronto Press, 2009.

[8] Volume 45 of Studies in Logic, London: College Publications, 2013.

Suffice it to say that if the view I oppose were to stand, then those who hold it would be left with no option but to accept ex falso's globalization of negation-inconsistent belief, *or* to put ex falso out of business, or to stop being so classically minded about belief. [9]

Inconsistency, of course, is a bad property; I mean negation-inconsistency is. But since negation-inconsistency globalizes itself, it makes of a bad thing a radically worse thing. Thus we see that inconsistency wreaks havoc in two directions. It destroys the good of good things and hyperinflates the bad of bad things.

8.1.3 Globalization Economics

The near-universal view is that inconsistency is bad for theories, and that an inconsistent logic simply can't get the job done. By a very wide margin, logicians loathe inconsistency like the plague. It stands to their theories as a spider to a damsel's petticoats. A goodly percentage of these same logicians harbour this hostility without express regard for whether ex falso is true. Some computer scientists of my acquaintance have never heard of ex falso, and some of these have had no express contact with the distinction between negation- and absolute inconsistency either. For them a single localized inconsistency is sufficient cause to abandon a theory in its present form—if not outright.

We value the beliefs furnished by our theories for the help they give us in advancing our knowledge of things. When all goes well, these systems play a large role in the cognitive economies of human agency. When globalization strikes, it wrecks the system, and it may also wreck its utility. The first spoilage is *environmental*. The second is *economic*. It is widely supposed that they go hand and hand. Actually, they don't. I will say why in due course. For the present let's go with the flow and assume that environmental wreckage entails economic collapse.

8.2 Damage Control

8.2.1 Damage Control

Reactions to inconsistency are numerous, varied, and rivalrous. They form multiplicities well beyond what is possible to cover here. But I will pause long enough to make a selective overview of some aspects of the leading treatments. I do so in ample recognition that, for a readership like this one, there isn't much in it that's new. But I do want to take note of how this taxonomy fatefully interacts with other contentions in the philosophy of logic.

The first broad partition divides responses that allow ex falso to stand from responses intent on bringing it down. In the former camp we have two prominent options. One is *preclusion*, which admits incoming information to a consistent system only upon a favourable consistency check. Its purpose is to keep inconsistency out. The other is *expulsion*. If preclusion sees trouble coming and then takes steps to avoid it, expulsion

---

[9] As James Joyce, for example, attempts to be. See his "The development of subjective Bayesianism", in Dov M. Gabbay, Stephan Hartmann and John Woods, editors, *Inductive Logic*, volume 10 of Gabbay and Woods, editors, *Handbook of the History of Logic*, pages 415-476, Amsterdam: North-Holland, 2010.

spots trouble upon arrival, and takes steps to remove it after the fact. We could say that expulsion is a putter of things right rather than an evader of things wrong. Despite their differences, there are difficulties they both share. Both require consistency checks; and yet for information systems of any realistic size—large enough to run the Five Eyes intelligence network, say, or to run the British Columbia health care system, or even to manage a household economy—consistency checks present problems of intractability. Even a truth functional consistency check is well beyond even the theoretical reach of beings like us. [10] Before going to the other side of our present distinction, it is appropriate to sound a warning. There are other distinctions of importance that cut across this one in telling ways. Until they are duly noted, we won't have an adequate grasp of it.

There is a substantial family of ex falso "resisters", sundered in turn by a further division. In the one camp we find ex falso "falsifiers". The second is peopled by ex falso "suppressers". *Ex falso* falsification requires a proof that ex falso is false, and that any proof of it is invalid as a matter of objective fact. *Ex falso* suppression requires its non-admittance. A system in which ex falso fails, whether by way of defeat or refused admittance, is a *paraconsisistent* system. [11] In its core sense, a logic is paraconsistent if ex falso is not provable there. [12] It is not required—though it could be true—that it also invalidates ex falso's proof. The main thing is to give ex falso no purchase. The falsification option was vigorously seized upon in the earlier days of relevant logic. Anderson and Belnap, Routley, Meyer and, later Read, would fashion arguments seeking to show that ex falso was objectively false and that what Lewis and Langford claimed to follow from their proof did not as a matter of fact follow at all. [13] But in the years

---

[10] It is known that the consistency problem is solvable in polynomial time by a nondeterministic Turing machine. They are also solvable for decidable logics by a deterministic computer in exponential time. But not by us. See, for example, Christopher Cherniak, *Minimal Rationality*, Cambridge, MA: MIT Press, 1986; esp. pp. 78-81, and Petr Hájek, *Metamathematics of Fuzzy Logic*, Dordrecht: Kluwer, 1998; chapter 5. The foundational papers for complexity theory include S. A. Cook, "The complexity of theorem-proving procedures", in Proceedings of the 3rd Annual ACM Symposium on Theory of Computing, pages 151-158, New York: *Association for Computing Machinery*, 1971, and "An overview of computational complexity", in *Communications of the Association for Computing Machinery*, pages 400-408, New York: Association for Computing Machinery, 1983. See also R. Karp, "Reducibility among combinatorial problems", in R. Miller and J. Thatcher, editors, *Complexity of Computer Computations*, pages 85-104, New York: Plenum Press, 1972. Also to be recommended is Cliff Hooker, *Philosophy of Complex Systems*, edit, a volume of D M. Gabbay, Paul Thagard and John Woods, editors, *Handbook of the Philosophy of Science*, Amsterdam: North-Holland, 2011.

[11] Priest, Routley and Norman, editors, *Paraconsistent Logic*, 1989; Bryson Brown, "Preservationism: A short history", in *The Many Valued and Nonmonotonic Turn in Logic*, volume 8 of Dov M. Gabbay and John Woods, editors, *Handbook of the History of Logic*, pages 95-128, Amsterdam: North-Holland, 2007; Graham Priest, "Paraconsistency and dialetheism", in Gabbay and Woods, volume 8 of *ibid.*, pages 129-204; Jean-Yves Béziau, Walter Carnelli and Dov Gabbay, editors, *Handbook of Paraconsistency*, volume 9 of Studies in Logic, London: College Publications, 2007; and Schotch, *et al., On Preserving*, 2009.

[12] Some paraconsistent logics admit ex falso. The idea seems to be that ex falso holds for entailment but doesn't call the shots for inference. As Gill Payette has reminded me there are preservationist systems in which fixed level forcing is complete for *n*ary modal logic. Such systems $\ulcorner A \& \sim A \vdash B \urcorner$ is valid, but $\ulcorner \{A, \sim A\} \vdash B \urcorner$ is not. They are non-adjunctive in the manner of Jaśkowski.

[13] Alan Ross Anderson and Nuel D. Belnap, Jr., *Entailment: The Logic of Relevance and Necessity*, volume 1, Princeton: Princeton University Press, 1975; volume 2 (also with J. Michael Dunn), 1995; John Woods, "Relevance", Logique et Analyse, 8 (1965), 211-220, and "On how not to invalidate disjunctive

8.2.2 A Trinity

In the ex falso-suppression camp,[14] there is yet a further distinction, separating *realists* from *relativists*, and each of these from *nihilists*. Nihilism is the doctrine that there are no ontic facts of the matter about the theorems and properties of a logical system. Relativism countenances logical facts as facts about a logical system. Realism asserts that logical facts are facts about the world. Nihilism thinks that logic leaves no metaphysical footprints of any kind. A relativist's facts leave metaphysical systems but not the world. A realist's facts leave metaphysical footprints on the world itself.

Of the three, nihilism stands out for its repulsiveness. Think here of Dana Scott's reaction to dialethic logic: (*Pornography*!) If there are no facts of the matter about the properties on which logicians routinely focus and for which they contrive their contributions, big trouble awaits. If there are no facts about following from, about validity, inconsistency and the like, then on the principle, *No entity without identity*, logic is denied a subject matter. There is nothing for logic to be about, and nothing for logic to get right (or wrong). Of course, nihilism lies open to a gentler interpretation. Some will see it as saying that there are indeed facts of the matter about following from, validity and the rest. But given the Babel's tower of conflicting answers, over-populated and lacking for successful conflict resolution strategies, the objectively correct story about logic's subject matter is simply unavailable to us. Imagine that! The world teems with a plethora of logical facts to which we poor souls have no reliable epistemic access. Call this "epistemic" rather than "ontic" nihilism. It comes at a high price. It is radical scepticism about matters that should be least vulnerable to it. It makes every logician's contribution little more than a hopeful shot in the dark.

A still lighter version of nihilism is *stipulationism*. Stipulationism accepts that concerning logic's subject matter there are no prior facts or at least none that is knowable. It accepts that in regard to that subject matter—and the properties and relations that make it up—there is no future. Stipulationism is a response to this impasse *faute de mieux*. Not only is it in such dire circumstances the better thing to do, it is the only thing to do short of a career in real estate or investment banking. According to stipulationism, there is no dearth of logical facts, indeed there is a veritable prosperity of them. But they are facts of a quite distinctive character. They aren't antecedently real objects. They are *objets d'art* not *objets trouvés*. They are facts of the theorist's own creative

---

syllogism", *Logique et Analyse* 8 (1965), 312-320; Robert K. Meyer, "Entailment" *Journal of Philosophy*, 68 (1971), 808-818; Alasdair Urquhart, "Semantics for relevant logics", *Journal of Symbolic Logic*, 37 (1972), 159-169; Robert K. Meyer and Richard Routley, "Dialectical logic and the consistency of the world", *Studies in Soviet Thought*, 16 (1975), 1-25; Stephen Read, *Relevant Logic*, op. cit; 1988, p. 22 ff; Richard Routley, "Relevantism, matrial detachment and the disjunctive syllogism argument", *Canadian Journal of Philosophy*, 14 (1984), 167-188; Richard Routley and Robert K. Meyer, "Relevance logics and their semantics remain viable and undamaged by Lewis's equivocation charge", *Topoi*, 2 (1983) 205-216; and John Woods, *Paradox and Paraconsistency: Conflict Resolution in the Abstract Sciences*, Cambridge: Cambridge University Press, 2003, pp. 58-68. My answer to Read can be found in *Errors of Reasoning*, pp. 525-6.

[14]Of course, ex falso falsifiers are themselves a proper subclass of ex falso suppressors.

efforts, as are the properties and relations with respect to which they are indeed the facts. (In Russell's usage, they are created by nominal definition).

Here is another peculiarity. The *vocabulary* of stipulationism is not made up, not inherently so at least; it is often a borrowed vocabulary from a failed enterprise. In prior usage, "follows from", "set", and the like are predicates having either null or unknowable extensions. They are good words with no useful work to do. Stipulationism lends them a helping hand. It offers new properties and relations for them to name. In its refreshed usage, "follows from" now has a briskly non-null extension, and similarly for "set" and all the others. But there is nothing whatever in those non-null extensions apart from items of the stipulationist's own makings-up. His following from and *his* sets are following from and sets in name only.

Here, too, finer distinctions are available for consideration, giving us a lattice of them. We might say that *actualist* stipulationism creates facts that leave new metaphysical footprints on the world; that *systems* stipulationism creates facts that leave new metaphysical footprints on theories; and that *nihilist* stipulationism creates no facts whatever, but makes up for it by creating fictions. The three are not strictly disjoint. A good historical example is Russell's treatment of sets after the paradox that bears his name. Kant would have called it "synthesis". Synthesis is the business of making new concepts out of pre-existing conceptual parts. It is not the business of making new concepts out of thin air. In Russell's hands, synthesis is a hybrid of pre-existing conceptual parts and wholly made-up ones, coming together by nominal definition in something genuinely new.

So far, preclusion strategies divide between those that refuse admittance to inconsistency and those others that allow for local inconsistency but refuse admittance to ex falso, with the intention of supressing the globalization effect. It bears repeating that the first of these is consistent with ex falso, while the second is a paraconsistent response. The third manoeuvre neither admits nor refuses entry to inconsistency, and takes a quite peculiar stand towards ex falso. On the one hand, it demolishes it utterly. On the other, it finds it wholly irrelevant. This is a view with few takers, but there are clear suggestions of it in Boethius and some ambiguous indications in Wittgenstein. [15] This, the *self-cancellation* view, asserts that with regard to any inconsistent conjunction $\ulcorner A \wedge \sim A \urcorner$, the conjuncts A and $\ulcorner \sim A \urcorner$ cancel one another out in ways that leave both of them with nothing to say, that is, shorne of propositional content. Since conjunction is undefined for non-propositions, nothing in the form $\ulcorner A \wedge \sim A \urcorner$ is a proposition, still less an inconsistent one.

This gives us three preclusion options with regard to $\ulcorner A \wedge \sim A \urcorner$. (1) Don't let it in. (2) Let it in but block further trouble. (3) There's nothing *to* let in or keep out.

---

[15] See L.M. de Rijk, editor, *Petrus Abealardus: Dialectica*, Assen: van Gorcum, 1970, p. 290; John Woods, "Dialectical considerations on the logic of contradiction I", *Logic Journal of IGPL*, 13 (2005), 231-260; and Ludwig Wittgenstein, *Remarks on the Foundations of Mathematics*, Oxford: Blackwell, 1964; p. 171.

8.2. DAMAGE CONTROL

8.2.3 Paraconsistency

With scarcely an exception, *ex falso* is paraconsistently dealt with by invalidation or suppression of a proof rule that supports it—for example, the DS rule. In falsification *milieu* DS is subject to a proof of ontic invalidity. In suppression *milieu* that aren't falsifiers, DS is simply denied a place in the logic. We see in this an interesting link to preclusion. As we saw it a page or so ago, preclusion was non-admittance of inconsistency. As we have it now, it can also be non-admittance of *ex falso* itself. There are by now more ways of being a paraconsistentist than could have been imagined in the late 1950s and 1960s, some of the most notable of which is the preservationism of Peter Schotch and his colleagues.[16] Even so, today's logics share some common features, apart from their agreement that *ex falso* will have to go. In many of these systems there are constraints on premiss-eligibility, especially in matters of joint use. In virtually all of them, the consequence relation is stripped down, re-jigged and retrofitted. The absolutely central complaint pressed by paraconsistentists is that any system admitting a consequence relation that supports *ex falso* has either got the consequence relation (ontically) wrong, or independently of that has extended its hospitality to a consequence relation that guarantees the system's environmental and economic devastation. If *ex falso* is true, inconsistency, however slight, goes global.[17]

It takes little reflection to see how odd this is. When Russell proved the inconsistency of set theory, no one thought to pin the blame on the consequence relation. Everyone knew that the blame fell on Frege's notion of set. When Tarski proved the paradox of the liar, the notion of truth was in trouble, not the following-from relation. This is standard diagnostic practice. When an inconsistency follows from something we used to hold dear, the fault lies with it, not with the following-from relation. To blame the consequence relation for our mismanagement of the turbulence embodied in our own misconceptions is Cassandrian spin with the nerve of a canal horse.

If we don't like *ex falso*, we'd do well seriously to consider that the fault, if there is one, lies elsewhere. It lies where the fault of the Russell and Tarski inconsistencies is sometimes thought to have lain. It lies in our own *understanding* of inconsistency.[18] So far as I know, there is only one branch of paraconsistent logic in which such reconsideration occurs and does so in a load-bearing way. The central message of dialethism

---

[16] An important early paper is Peter Schotch and R. E. Jennings, "Modal logic and the theory of modal aggregation", *Philosophia* 9 (1981), 265-278.

[17] See for example, Stanis?aw Jaśkowski, "Propositional calculus for contradictory deductive systems" *Studia Logica*, 24 (1969), 143-157; first published in Polish in 1948; Nicholas Rescher and Ruth Manor, "On inference from inconsistent premisses", *Theory and Decision*, 1 (1970-71), 179-217; Nicholas Rescher and Robert Brandom, *The Logic of Inconsistency*, Oxford: Blackwell, 1980; Schotch and Jennings, "On detonating", 1989; Bryson Brown, "On paraconsistency", in Dale Jacquette, editor, *A Companion to Philosophical Logic*, pages 628-650, Oxford: Blackwell, 2002; Diderik Batens, Chris Mortensen, Jean Paul van Bendegem, and Graham Priest, editors, *Frontiers of Paraconsistency*, Baldock, Herts: Research Studies Press, 2000; Batens, "A survey of inconsistent-adaptive logics", in Batens et al. (2000), pages 49-73; and Graham Priest, "Paraconsistent belief revision", *Theoria*, 68 (2001), 214-228; revised as chapter 8 of Priest, *Doubt Truth to Be a Liar*, New York: Oxford University Press, 2006.

[18] In my own view, this is only partly right. It lies in what *occasions* our misunderstanding of inconsistency. See here my "Does changing the subject from A to B really enlarge our understanding of A?", *Logic Journal of IGPL*, 2016, 1-25; online doi:10-1093/jigpal/jzw017.

is that we've got the concept of inconsistency wrong. We've got it wrong in precisely those cases, and some few others like it, in which concepts embodied in the premises have been fundamentally misconceived. Dialethism proposes a radically different diagnosis of these paradoxical inconsistencies. In so doing, it takes the view that while the paradoxical proofs are valid, their respective conclusions aren't false after all, still less absurd. In each case, the conclusion is a sentence in the form $\ulcorner A \wedge \sim A \urcorner$. Virtually the whole world is at one in thinking that the nature of the inconsistency resides in the fact that $\ulcorner A \wedge \sim A \urcorner$ takes F for all values of A.

8.2.4 Dialethism

Against this dialethists[19] propose that this is the wrong way to understand inconsistency when instantiated by the Russell and Tarski sentences. The right way, they say, requires a many valued approach to the semantics of those contexts. We now have four truth values, not two T, F, (B (for both) and N (for none). That alone tells us something about inconsistency. It tells us that a two valued semantics will get inconsistency wrong. In a functional Routley-star semantics for a suitably extended system of first-degree entailment, FDE B is the *set* $\{T, F\}$ whose members are just the prior two, which in turn are undefined semantic *objects*. As with the two valued case, the truth values of dialethic logic are mutually exclusive and jointly exhaustive.[20] No Routley-star sentence or sentence of LP has more than one of these four. (So, for example, having $\{T, F\}$ is not having T and having F.) Consider now $\ulcorner A \wedge \sim A \urcorner$ when they are the Russell and Tarski contradictions. They are not T and F; rather they are $\{T, F\}$, hence neither T nor F. The same applies to the Russell and Tarski conjuncts. If they were T or F, they would be T or F in the conjunction, and the conjunction would be F, hence not $\{T, F\}$. So when the conjunction *is* $\{T, F\}$ so too are its conjuncts.[21]

Much clamour has been produced by the claim that a dialethicized Lewis-Langford proof won't go though for *ex falso* in Routley star systems. It fails in its terminal line because of the failure of DS on the preceding two. If A is $\{T, F\}$, and $\ulcorner \sim A \urcorner$ is $\{T, F\}$ then $\ulcorner \sim A \urcorner$ and $\ulcorner A \vee B \urcorner$ together can't do the rule's business, which is predicated on $\sim A$ and A's having *conflicting* truth values. But there is no such conflict here. There is no difference in truth value.

---

[19]The coiners of this term spell it "dialetheism". Its component parts are "*dia*" and "*aletheia*", the first meaning straight through, in different directions" and the second meaning "truth". This would seem to give "*dialetheia*" with the extra "e" in. But the adjective "*alēthēs*" gives "alethic", with the extra "e" dropped out. Either usage is defensible, but "dialethic" is better in adjectival form; but since Greek abstract nouns hang on to their "e"s, "dialetheism" might strike the eye and ear more kindly. Perhaps the better course is to anglicize the lot and leave Greek grammar to grammarians. I am grateful to my colleague Michael Griffin for helpful instruction on these matters. It improves upon the somewhat less good observation to this same effect on page 113 of John P. Burgess, *Philosophical Logic*, Princeton: Princeton University Press, 2009.

[20] Graham Priest, *An Introduction to Non-Classical Logic: From If to Is*, 2nd edition, New York: Cambridge University Press, 2008; chapter 8, and "The logic of paradox", *Journal of Philosophical Logic*, 8 (1979), 219-241. The system therein is known as LP.

[21]Dialethicized FDE is also subject to a relational semantics in which B is both T and F. There are serious problems with this, I think. For more see "Does changing the subject ?". Either it can't block the proof of *ex falso* or the "and" of "T and F" isn't a sentential connective.

8.2. DAMAGE CONTROL

It strikes me as misconceived to attach so much of dialethisms's significance to its blockage of a dialethicized Lewis and Langford proof. It is just another case of *ex falso* preclusion. It is a preclusion that causes no grief for the consequence relation. It really doesn't follow from ⌜$A \vee B$⌝ together with ⌜$\sim A$ urcorner that B, when A and ⌜$\sim A$⌝ are identically valued. This leaves it entirely open as to whether it really does follow from ⌜$A \vee B$⌝ and ⌜$\sim A$⌝ that B, when A and $\sim A$ denote T and F respectively, or otherwise F and T. What is at issue here is not what follows from what. What counts is the way we conceive inconsistency.

It is not my purpose here to take up the broader question of whether dialethism has got inconsistency right. I'll simply make do with a further two observations. Even if dialethism were the right way to understand the inconsistency of the Russell and Tarski sentences, it is certainly by its own lights not the way to understand nearly all of what remains in the domain of inconsistent conjunction. Dialethists oppose near-trivialism with at least the vigour with which they resist trivialism. So, for all those other conjunctions ⌜$A \wedge \sim A$⌝, inconsistency lies as usual in the truth value incompatibility of A and ⌜$\sim A$⌝, for which the only available suppression option is one or other variation of paraconsistency. Let me say it again: it is no trick at all to show that it really is the case that not everything follows from a dialethically interpreted inconsistency. It is something quite different to show that not everything whatever follows from a two-valuedly interpreted inconsistency. For those who are contrarily minded, the standard options lie ready to hand: falsify *ex falso*; deny it admittance; or summon up relativism or nihilism. Whichever the better of these options might be, dialethists deserve substantial credit for having—if only in the limited domain of *paradoxical* inconsistency—shifted the weight of discussion from the nature of the consequence relation to the nature of the inconsistency property and to how it might sensibly be crisis-managed.[22]

My second observation is that if the two valued interpretation is right for inconsistencies in their transfinite plenitude, minus the handful that call for dialethic construal, two additional questions press for attention. One is whether a principled basis is available on which to manage the distinction between two-valued inconsistency and multi-valued inconsistency. The other is how, if the two-valued interpretation is right for all inconsistencies or all except for the meagre dialethic sprinkling of them, and if the real point of the *ex falso* fuss is whether we've got inconsistency right rather than doing the same for the consequence relation, then what justifies our blowing the whistle on how we understand consequence? Why wouldn't we gang up on how we understand inconsistency? Why wouldn't we see that giving up on consequence is a counsel of desperation, tied to the heavy onus of finding independently discrediting considerations? And what, pray, would those independently discrediting considerations be? We should also look for independently disabling considerations regarding two valued inconsistency. And here, too, I want to ask what would those considerations would be? In neither case have

---

[22] The logicians of Halifax, Vancouver and Lethbridge are crisis-managers *par excellence*. So are the paraconsistent logicians of the southern hemisphere. Priest's own crisis-management is underlain by a deep attachment to metaphysics in the grand manner. I refer the doubtful to his *One: Being an Investigation into the Unity of Reality and its Parts, including the Singular Object which is Nothingness*, New York: Oxford University Press, 2014.

I the foggiest idea.

Not even dialethists think that a dialethic construal is right for inconsistency in the general case. They think that dialethic construal is nearly always wrong for inconsistency. We could, if we liked, concede them the option of a purpose-built logic to accommodate the precious few that have caught their attention. Of course, for this they will need to block *ex falso*, but their dialethism doesn't require that they be paraconsistents across the board. That they *are* paraconsistents across the board was a separate decision. For as we saw, the collapse of the Lewis-Langford proof for three-valued inconsistency doesn't lay a glove on the proof for two-valued inconsistency.

There is, I say, no good independent reason to think we've got inconsistency wrong, and no good independent reason to think that we've got consequence wrong. If properly conceived inconsistency and properly conceived consequence jointly secure the truth of *ex falso*, why wouldn't we just face up to it and see that it lies in the nature of inconsistency that everything does indeed follow from it. Why wouldn't we concede that the Lewis-Langford proof is a sound demonstration of a surprising truth about inconsistency, perhaps even a shocking and disappointing one, as many of us appear to think?

The answer—to the extent that there is one—lies in what inconsistency does to theories and belief systems. It lies in the globalization of their hitherto desirable properties. It lies in the ensuing epistemic paralyses. This gives us options to consider. As we have it now, most favoured is that we've got consequence wrong. A distant second is that we've got inconsistency wrong. There is a third that seems not to have shown up on logic's radar. It is that we've got *globalization* wrong, that we've been wrong to think of globalization as epistemically paralyzing.

8.2.5 Schematic Summary of the Foregoing

It might be handy to have for ready reference some summary grasp of the damage control measures we've been discussing in the past several pages. We start with a damage control map, beginning with control-measures that leave the question of *ex falso*'s truth unengaged, then passing to those for which *ex falso* is the central concern. We conclude with a map of the conceptual and philosophical rivalries that intersect

8.2. DAMAGE CONTROL

with, and add to the complexity of, the damage control maps (see figure 8.1).[23,24]

A further consideration I won't take the time to diagram is logic's *multiplicity problem*. The key idea is this. Three facts about logic today stand out. Never has it been done with such technical virtuosity. Never has there been so much of it. Never has there been so little consensus about its shared subject matters. It would seem that the more we have of it, the less our inclination to get to the bottom of its sprawlingly incompatible multiplicities. There is nothing like this in real analysis, physical chemistry or population genetics. Left undealt with, one might see in logic's indifference to its own rivalries some sign of not quite knowing its own mind.

Most disciplines have conflict resolution measures to subdue their own contentious theoretical multiplicities. Untamed multiplicity is something of a scandal in physics— as witness the angst produced by the tension between the Standard Model and the shifting fortunes of string theory. There are two things troubling about logic's own shared subject-matter abundances. One is that they exist. The other is that hardly any working logician pays them much mind. Even so, there is a philosophical minority that does worry about these things. The most interesting of these is *pluralism*. Pluralism is an attempt to give to common subject matter multiplicities something of a good name. There is no time for this now, except to note that at bottom pluralism is a kind of ambiguation strategy. The heart and soul of it is to deny the commonness of the subject matters around which these multiplicities pile up. I will only say here that pluralism forgives the fact that there are more logics of the consequence relation than you can shake a stick at, by postulating more *consequence relations* than you can shake a stick at. Thus does one multiplicity wash the laundry of another.[25]

---

[23]For somewhat-welcoming treatments of inconsistency, see Dov M. Gabbay and Anthony Hunter, "Making inconsistency respectable 1: A logical framework for inconsistency in reasoning", in Ph. Jourand, and T. Kaleman, editors, *Fundamentals of Artificial Intelligence Research*, volume 535 of *Lecture Notes in Computer Science*, pages 19-32, Berlin: Springer, 1991, and "Making inconsistency respectable part 2: Meta-level handling of inconsistency", in M. Clark, R. Kruse and S. Seraffin, editors, *Lecture Notes on Computer Science*, volume 747, pages 129-136, Berlin: Springer, 1992; Anthony Hunter, "Reasoning with contradictory information using quasi-classical logic" *Journal of Logic and Computation* 10 (2000), 677-703; and Lepoldo Bertossi, Anthony Hunter, and Torsten Schaub, editors, *Inconsistency Tolerance*, Lecture Notes in Computer Science 3300: A State-of-the-Art Survey, Berlin: Springer 2004. For logics that treat of inconsistency-robust information systems, see Carl Hewitt, "Inconsistency robustness in foundations: Mathematics proves its own consistency and other matters", and "Formalizing common sense for inconsistency-robust information integration using Direct Logic reasoning and the Actor model", both in Hewitt and Woods, editors, *Inconsistency Robustness*, volume 52 of Studies in Logic, London: College Publication Publications, 2015, at pages 104-157 and 3-103 respectively; and also Woods, "Inconsistency: Its present impacts and future prospects", pages 158-194 therein.

[24]The standard paraconsistentist take on dialethism is that it doesn't bear thinking about. The standard preservationist take on that take is that "dialethism, rather than fixing truth-preservation breaks it beyond any hope of repair". (Schotch, *On Preserving*, p. 12) However in his "Ambiguity games and preserving ambiguity measures" (*On Preserving*, chapter 10), Bryson Brown produces "a way of connecting preservationism with inconsistency-dialethism. This is rather startling to say the least." (Schotch, ibid, p. 14)

[25] See, for example, JC Beall and Greg Restall, *Logical Pluralism*, New York: Oxford University Press, 2006; Hartry Field, "Pluralism in logic", *Review of Symbolic Logic*, 2 (2009), 342-359; and John Woods, "MacColl's elusive pluralism", in Amirouche Moktefi and Stephen Read, editors, *Hugh MacColl After One*

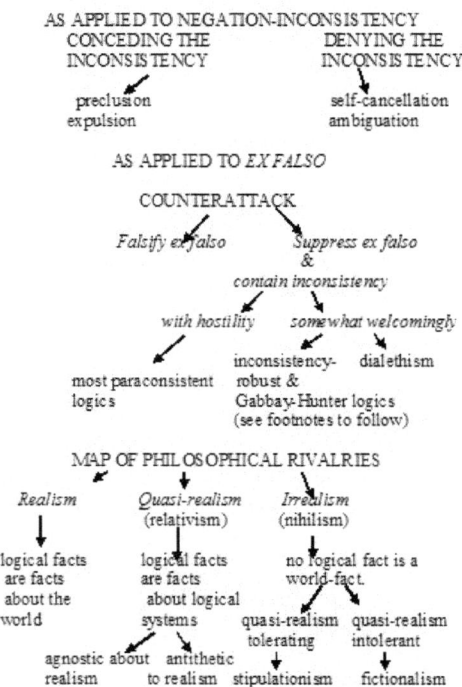

Figure 8.1: Damage Control Map

Coming back to the rivalries map, here too is a taxonomy that divides the current literature in multivarious and ways. As I write, each of its elements poses an open question for logical theory—which in the main is either ignored by the logic's system-builders or closed by stipulation. This is explicable if you're a died-in-the-wool irrealist. If you are an irrealist, there is no fact of the matter that closes these questions. But that is big-box scepticism. It is scepticism on wheels. Speaking for myself, big-box scepticism in logic, of all places, is an intellectual set-back of high order.[26]

In strictly operational terms no inconsistency-management policy can play out independently of these background considerations. At every joint of the damage control map, they occasion and motivate further refinements and qualifications. Both maps

---

Hundred Years, pages 205-233, Paris: Editions Kimé, 2011; a guest-edited volume of *Philosophia Scientiae*, 15, 2011.

[26] For an interesting take on scepticism, see Peter Schotch, "Skepticism and epistemic logic", *Studia Logica*, 65 (2000), 187-198.

8.3. UNRECOGNIZABILITY

trace unresolved conceptual and methodological turbulence. What is required is for the two turbulences to converge on genuinely productive conflict resolution strategies for the stabilization of logic and the management of inconsistency. On the face of it, it is an overwhelming task, threatening any would-be completer of it with the fate of man-overboard. Unless, of course, we can think of some means of escape.

8.3 Unrecognizability

8.3.1 The Economic Utility of Wrecked Environments

For any information or belief system which harbours negation-inconsistency, none of its containment strategies stands any realistic chance if *ex falso* is true. Paraconsistentism and dialethism would be wiped out in one way or another. Having lost any semblance of realistic legitimacy, their only retreat would be a relativism that does them no credit. Consider again the case of set theory. We want a theory of sets to tell us what's true of sets and what not. We want a theory of sets to enlarge our knowledge of sets beyond what we already knew going in. Naïve set theory was like this too. We wanted it to tell us what the set-theoretic theorems are. Of course, naïve set theory embeds the Russell inconsistency. If *ex falso* is true, this occasions environmental devastation, inflicted by the globalization of provability. It would appear to be an economic devastation as well. If we wanted the theorems of naïve set theory to advance our knowledge of sets, those with stabilizing transfinite arithmetic, we wouldn't want to know that all its sentences are theorems and yet have provable negations. Naïve set theory seems powerless to advance our cognitive interests. It suffers from a paralyzing epistemic impotence. Wouldn't this imply that the environmental blight of naïve set theory converts straightaway to economic collapse?

In what remains of this paper, I'll show that this is not the case. My argument will be advanced in two stages, one forwarding a logical fact, and the other registering some empirical observations. In the first, I'll argue that *ex falso* is demonstrably true, that it is a fact that leaves metaphysical footprints on the world. In the second part, I'll present considerations showing that economic wreckage thesis is untenable as a matter of empirical fact.

8.3.2 A Logical Fact

Let's come back to the proof of *ex falso*. I take a realist approach to *ex falso*. But my method of proving it is not quite Lewis's and Langford's way. As we saw in section 2, lots of people judge the Lewis-Langford proof to be defective, not because it derives from a misunderstanding of inconsistency, but rather because of its reliance on defective connectives, whose truth functionality rendered them unfit for the job at hand. Be that as it may, I want to avoid those connectives. I also want to avoid a proof constructed for an uninterpreted formal language. What I want is a proof that is deliverable in English, without the aid of "or" in its use as a connective of English. It is the proof I offered in section 8.1.2, with apologies to readers who disdain repetition.

1. Assume for argument's sake that S and not-S, for some statement-expressing sentence S of English.

2. Then on the principle that if both are assumed then each follows, we have it that S.

3. By the principle that if some one thing follows, so too does at least one of the pair made up of it and anything else, at least one of $\{S, X\}$ follows for arbitrary X.

4. By the reasoning that took us from line (1) to S, we now go to "not-S".

5. By the principle that if at least one of two sentences follows, and it's not this one, then it is the other one, we have it now that X by the negation rule.

If, as I will now say, each step of the argument is truth-preserving, and if "S and not-S" is assumed, there is a truth-preserving route to any X whatever. And, approvingly in the spirit of Lewis and Langford in their behalf, I ask critics to show in a non-question-begging way where this argument really does misfire? I ask it with rhetorical intent, but I am prepared to be snapped out of it by a convincing answer.

Note, by the way, that the proof rules of my version of the Lewis-Langford proof are not syntactic rules in the present-day sense. It may be that they bear some similarity to syntactic proof rules. But it remains an open question as to whether those rules formally represent my own. We should also bear in mind that, what my argument doesn't show. It doesn't show that *ex falso* holds *relative to English*. No natural language has the relativity property that formal systems have. Irrespective of whether it actually holds, *ex falso* does hold relative to classical logic. It also holds relative to the S-systems of modal logic and many modal others besides. It fails relative to lots of other systems too. It fails relative to the basic relevant system R and it fails in the dialectic system LP and Routley-star FDE. It fails relative to virtually all other paraconsistent systems. There is a sense in which in all these cases *ex falso* holds or fails on the authority of the system. This is not the way with English. A proof of *ex falso* is formulable in English, but it doesn't hold on the authority of English. It holds on the authority of the world. Let me say a brief further word about this. The human animal, among others, is a *can't-help-it realist*. When we have an onrushing-tiger-experience, there is adaptative advantage in seeing that a tiger rushes on. When we have a throat-torn-out experience, there is no advantage *now* in knowing that a tiger is relieving us of our throats. But there is an inherited adaptation. It reinforces expectations at large that when onrushing experiences occur throats are at significant risk. All natural languages, whatever their capacity for nuance or equivocation, are nature-built to accommodate can't-help-it realism. Reification and individuation are the hallmarks of human languages and the structure of humanity's contact with a menacing world. If can't-help-it realism weren't adaptively advantagous—something pretty close to as natural as breathing there'd be no one to pay tribute to Peter. No Peter either.

Since time immemorial, sages have aligned their forces against can't-help-it presumptions, receiving a hefty endorsement from the family of scepticisms loosed by the Attic revolution of *logos*. When Gorgias went home after a destructively productive day at the office, he would gather himself and set out for the week's favoured symposium in

8.3. UNRECOGNIZABILITY

precisely the same basic frame of mind as the slave boy who carried his belongings. When Gorgias wrote that nothing exists, or if it does it can't be known, or if it can it can't be communicated, he said these things as fact, and said so in a language equipped by nature to serve the demands of can't-help-it realism. There are those, lots of them, who think that can't-help-it realism is delusional, never mind its "instrumental value". What are they thinking in so saying? If can't-help-it realism is delusional, so is the idea that it's instrumentally valuable. That's the trouble with big-box scepticism. It is not a pragmatically possible doctrine to proclaim or even formulate in the mind of a human being.

Back to my logical fact. The logical fact I want to advance is that negation-inconsistency really does globalize itself. What now of the empirical observations?

8.3.3 Some Empirical Facts

The empirical facts are these. The Newton-Leibniz calculus was inconsistent. If *ex falso* were true, calculus would have been a globalized wreck. There would have been nothing to be learned from it, whether about the infinitesmally small, or integration and differentiation or anything else. If *ex falso* were true, calculus would be a system supersaturated with noise. There is no working mathematician who believes this to be so. There is no working physicist who thinks that physics was a train-wreck in the two centuries that awaited Weirerstrasse's limits or (later) Robinson's hyperreals. The question in view is whether environmental wreckage implies economic wreckage and the cognitive paralysis that attends it. The empirical fact is that no one in the world-wide, centuries-long community of working mathematical physicists had any time for the idea that there was nothing to learn of cosmology from Newton's *Principia*. Their entirely settled practice massively discomports with the thesis that environmental damage destroys economic advantage.

The chief prosecutor of the inconsistency charge was Berkeley in his 1734 monograph *The Analyst*. Neither Newton nor Leibniz was much disturbed by it. Newton knew of a method for getting rid of the inconsistency but never got around to publishing it until long after *Principia*'s appearance in 1687. (Newton's later provisions were a good adumbration of the method of limits). Leibniz denied a real commitment to infinitesmals and seems to have used them only as handy fictions. A perhaps more definitive case of inconsistent science is Bohr's theory of the atom, from whose postulates together with Ehrenfest's adiabatic principle, Wolfgang Pauli derived a flat-out contradiction. Newton, we see, was a laggard expulsionist and Leibniz a pragmatically oriented fictionalist about infinitesmals, or anyhow a system-relative quasi-realist. But neither would have denied that in the form of its publication in 1687, the *Principia* is indeed inconsistent. The Bohr model also had empirical shortcomings. Since Bohr was a quasi-realist about physics, he would have been less bothered by his model's inconsistency than by its empirical shortfalls. [27]

---

[27] These and related matters are thoroughly examined in Peter Vickers' *Understanding Inconsistent Science*, New York: Oxford University Press, 2014. Of particular interest is Vickers' claim that reports of the inconsistency in science have been greatly overblown. If by inconsistency we mean the derivability of a contradiction from a finite subset of a theory's sentences, then it simply isn't true, says Vickers, that

Consider now the case of Reggie Classicus. Dr. Classicus is a mathematics professor who has written a one-semester introduction to elementary set theory called *An Introduction to Elementary Set Theory* suitable for use by high school and first-year university students. The aim of the text is to supply students with enough of a command of sets to arrive at a workable understanding of how sets hook up with the transfinite. Based on the early Frege axioms, the course divides into two parts. Part one covers the basic material about sets—membership, inter-set relations, operations on sets, and so on. Part two makes the transfinite connection at a level of rigour typified by the Cantor-Berstein-Schröder theorem. The underlying logic of the book is the standard functional calculus of predicates with identity. (Dr. Classicus is well-named.)

Towards the end of part one, the Russell paradox is derived and informally discussed. Experience with the text testifies to a lively reception of this news, but no measurable apprehension over it. It is explained that since the paradox is beyond the scope of the course, the original axioms will be allowed to stand. Moreover, Dr. Classicus doesn't trouble to acquaint his readers with ex falso *quodlibet*. We may therefore take it that while students are aware of the negation-inconsistency of the theory they're being taught, they aren't (in the general case at least) aware of its absolute inconsistency. Dr. Classicus is another story. He is a classical logician and a set theoretic originalist. He cannot not know that his theory has gone global. And yet he was able to tell his readers enough about sets to support the level of sophistication required for the Cantor-Bernstein-Schröder. He was able to make a principled connection with the transfinite. Who, again, would say that there was nothing for his readers to have learned from this globalized mess? How likely is Classicus to be indicted for fraud, or cashiered from his university for incompetence? How can we not say that here is a case of excellent instruction—a real advancement of learning—conveyed by a theory that is known to be polluted?

My suggestion is that environmentally wrecked systems—systems whose virtues are crushed by globalization—are capable of fruitful employment as instruments of cognitive and practical advantage. This leaves us with a good and necessary question. How come?

8.4 How Come?

8.4.1 Tracking Truth

I don't really know. But some of the facts appear to be these. They tell us something important about theorems. Theorems, in the syntactic sense that engages us here, aren't (intrinsically) truth-tracking. $\ulcorner \sim A$ is a theorem$\urcorner$ needn't contradict $\ulcorner A$ is true$\urcorner$. Every

---

inconsistencies plague science systematically. Others are otherwise minded. In *Paraconsistent Logic*, we read that inconsistencies in science abound. (p. 152) Not only is the Newton-Leibniz calculus inconsistent, Newton's theory of gravitation is inconsistent with Galileo's law of free fall and Kepler's laws. Statistical thermodynamics is inconsistent with second law. Wave optics is inconsistent with geometrical optics. Bohr's model of the atom is inconsistent with Maxwell's equations. The special theory of relativity was persisted with notwithstanding its inconsistency with the Kaufmann experimental results in 1906. However, with the exception of the inconsistency of the calculus, and short of their unification in amalgamating theories, none of the other cases meets the Vickers' test.

8.4. HOW COME?

sentence of naive set theory is a theorem, but theoremhood is neither truth-tracking nor belief-inducing. Of course, some of Dr. Classicus' theorems are indeed true. Some induce belief, and others don't. The working mathematician has a good, if not perfect, record in believing mathematical truths and not believing mathematical falsehoods, even in the absence of proof. Good mathematicians have a good nose for truth. Great mathematicians have a nose for unproven deep truths. Think here of Fermat.

Mathematicians are interested in propositions they believe to be true. Except conditionally or for *reduction*, they do not apply themselves to propositions they believe to be untrue, even when their proof rules prove them to be theorems. Accordingly "Everything is a theorem" has no premissory presence in Classicus' truth-tracking proofs, still less the Russell sentence or the Liar. Perhaps we will see in Russell's own response to his paradox some useful instruction. In *Principles of Mathematics* (1903), Russell asserts flat out that the concept of set had been proved insusceptible of a philosophical analysis. He went on to say that there was no way in which to do that concept's intended work in the foundations of arithmetic, short of nominal definitions by which something new would be constructed stipulatively. Sets would be entirely made-up mathematical entities. One might easily think that what Russell was trying to do for a concept that cannot be philosophically analyzed was provide it with a mathematical definition. In this transition from philosophical inarticulability to mathematical articulability, we might think that Russell see one and the same concept in transit, the concept that's philosophically unanalyzable and yet which is nevertheless susceptible of nominal definition. On reflection, however, I think that this cannot have been Russell's view, which proposed something radically different. It proposed that there was no concept at all that made this journey from set theory in 1893 and to set theory after 1902. Russell, like Frege, thought that what the paradox had revealed is that there isn't, and never was or would be, a concept of set at all. The difference between Frege and Russell was that Frege thought this an irrecoverable loss, whereas Russell got to work cobbling together a new one.[28] Certainly it would have been Frege's view that there is nothing of the old concept in Russell's made up one. It appears to me that this would also have been Russell's view. The likelier hypothesis, and the one that's especially instructive here, is that while *Grundgesetze* I afforded Russell no *conceptual* guidance in his post-paradox quest *to make* things up, there were its pre-paradox *descriptions*, which, though wholly non-referential, are perfectly intelligible sentences of German. Taken collectively they tell a story of nothing. The trick now is to contrive something anew for those descriptions, some of them anyhow, to be true *of*. Both Russell and Frege had a decidedly low opinion of the philosophical relevance of fiction, even in relation to the philosophy of natural language. That was decidedly silly of them, but not even they thought that the stories of Homer and Arthur Conan Doyle were incoherent or unintelligible or entirely without virtue. But their virtues, whatever else they might be, aren't philosophically analytical ones; they are instrumental ones. By 1903, Russell's own instrumental pragmatism was vigorously in view, and in due course

---

[28] Bertrand Russell, *Principles of Mathematics*, 3rd edition, London: Allen and Unwin, 1937, first published in 1903; pp. 15, 27 and 122 in the 1937 edition. See also Woods, *Paradox and Paraconsistency*, pp. 137 ff.

he would develop an affection for the "logical fictions" that would guide our future philosophical advances and spare us our metaphysical and epistemological pitfalls.

I see no reason not to think the same of Classicus, and of the Newtons, Leibnizs, and Bohrs before him. Although inconsistent theories cannot possibly be true, still less contradiction-inducing axioms, they can serve as stories in quest of something to be true of upon suitable pruning. In the Classicus case, they are those of his over-plentiful theorems that are truth-tracking. The theorems don't prove the truth of the truths they track to. But without their instrumental guidance, some of those truths might have indefinitely eluded us.

It shouldn't be all that surprising then that people who are good at mathematics would also be good at advancing our knowledge of sets even in environments in which syntactic theoremhood has gone viral.

Classicus' nose for the theorems of set theory that are actually true strikes me as similar to our own nose for hypothesis-selection. Peirce is good on this. He takes it as given that science couldn't progress if scientists weren't good at hypothesis-selection. The plain fact is that good scientists are indeed good at hypothesis-selection. Peirce puts this down to good guessing. Guessing, he says, is a trait common to us all, and our adeptness at it instinctive and innately grounded. I don't know whether this is literally so, but it does certainly seem to be the case that if we weren't guessers, we'd have perished long since, and if we weren't good at it, we'd also be dead. Good guesses are not inherently truth-tracking, but some of them are true. Without a secure sufficiency of true guesses, there'd be no science.[29]

The thesis that environmental wreckage implies economic wreckage turns out to have been a striking and consequential mistake. It is also a revealing mistake. It helps us see that cognitive prosperity is achievable with the aid of the axioms and theorems of wrecked systems. It tells us that, while inconsistency globalizes itself and provability too, it neither globalizes nor kills knowledge-acquisition. It tells us that from the perspective of knowledge globalization makes inconsistency productively unrecognizable.

8.4.2 Soundness

. I imagine that by now some readers will have grown impatient with the musings of the preceding section. "Doesn't he know", they might be wondering, "that the whole point of *sound* logistic systems is to show that their theorems are indeed truth-tracking?" So it is, and when that purpose is achieved, a system's theorems stand in a tight extensional equivalence to its *own* logical truths. Let L be such a system. Its theorems are those L-sentences that are provable in L. L's logical truths are precisely those L-sentences that have a model in every interpretation of L. If, for example, L is a first order theory, we could say that its logical truths are just those sentences satisfied in all interpretations by every countably infinite sequence of elements from L's set theoretic domain. It is true that when a system such as L is taken as a formal representation of an intuitive theory, the intention of L's makers is to have its logical truths pair one-to-one with a

---

[29] Charles S. Peirce, *Collected Papers*, eight volumes, Cambridge, MA: Harvard University Press, 1931-1958; 5.171 and 7. 220, and Peirce *Reasoning and the Logic of Things: The Cambridge Conference Lectures of 1898*, edited by K. L. Kettner, Cambridge, MA: Harvard University Press, 1992; p. 128.

8.4. HOW COME?

subset of the natural language truths of the represented intuitive theory. In some cases, that intention is thought to have been realized. The theorems of formalize arithmetic are thought to pair one-to-one with the truths of Peano arithmetic. In other cases, it is clear that no such intention is even in play. Logicians of irrealist stripe have no designs upon the pre-existing truths of some or other intuitive theory. In their approach, the system's target concepts are made-up ones, and its theorems and logical truths are in *their* service. On some fair readings of Hilbert, this was precisely his own intention. In its "pure" version his goal was to break once and for all the habit of mathematicians to tether its theorems to mathematical truth, and to allow the totality of logic's ontically cleansed theorems to make them meaningful. But by and large, the working mathematician is not in the business of just making things up, except as instrumentally advantageous in the revelation of the further pre-existing truths of his subject matter. The same holds for such propensity as the working mathematician may have for formal representability or formal modelling. His focus throughout is the light his system sheds on the truths of the unformalized discipline which is the working mathematician's home turf.[30]

There are formal systems galore for which the desire to have logical truths pair with the intuitive truths of a subject matter would be forlornly misbegotten. Examples come to mind. Lewis' S3 is one of them, and Halldén's S7 another. There is (say I) no sense of "necessary" and "possible" in which the logical truths of a sound and complete formal semantics for S3 or S7 pair one-to-one with intuitive truths about necessity and possibility in any pre-existing sense of these words.[31] Notoriously, no logical truth of S3 or S7 pairs with the intuitive claims that $\ulcorner S \vee \sim S \urcorner$ is necessarily necessary and its negation necessarily impossible. It is certainly the case that the logical truths of an counterintuitive (e.g. purely stipulated) subject matter sometimes can be made to stand one-to-one with the system's syntactically drivable sentences. But if it is the antecedently existing truths of a subject matter that *motivate* the modeller, he will be wasting his time with formal arrangements whose logical truths fail to deliver the intuitive goods. When a number theorist wants to know what number theory is true of, he wants to know what's true about *numbers*, not about what has a *model* in every interpretation of some formal system. What he wants from natural number theory is conclusive exposure of those actual truths. If he is trying to systematize those disclosures in a framework that turns out to make everything provable about numbers, that intention is ill-served by that framework. But as I think our reflections of late convincingly show, though the theory is disabled for fulfillment of its founding intention, this fact alone need not erase the pedagogical good that is in it.

8.4.3 Recognizable Presence

Of course, not everyone believes *ex falso*. Most people have no command of the question it answers. It is perfectly explicable that, even in the presence of what logicians

---

[30]There is more to be said about this, but not here. Here, too, interested readers could check out "Does changing the subject from A to B really enlarge our understanding of A?".

[31] Andrew Irvine, for one, thinks otherwise. He thinks that, although irregular and non-normal, S7 gives a better account of the philosopher's notion of possibility than do S4 or S5. See his "S7", *Journal of Applied Logic*, 11 (2013), 523-529.

know as negation-inconsistency, people wouldn't tumble into the chaos of cognitive dissonance to which the truth of *ex falso* would appear to condemn them. "What you don't know won't hurt you", as the old saying goes. In cases such as these, it hardly needs saying that unrecognizability is a matter of course.

Unrecognizability is a lot more interesting for anyone who actually believes *ex falso*, including the bulk of the world's mainstream logicians. In their case, globalization is known and acquiesced to. Where is the sense in saying of these folk that absolute inconsistency is unrecognizable to them? How can they not recognize the very thing that they believe to be true? One would think that the only alternative left to them would be Frege's. Faced with the Russell paradox, Frege came to the view that there is nothing whatever to *be* known about sets, other than nothing *is* one. This occasioned his permanent retirement from the philosophy of arithmetic. To the best of my knowledge, no other classically minded logician has followed his example. Why, then, are they still in business? The answer, in part at least, is that the globalizing inconsistencies they believe to be present in the work that they do fail to induce the very cognitive dissonance that destroyed Frege's set theoretic ambitions and his own capacity for their realization. Unrecognizability in this case has a character all its own. It is the cognitively composed acknowledgement of environmental wreckage, with virtually no alterations to business as usual.[32]

Here is a further point about unrecognizability. Questions from the floor of the House of Commons are not permitted unless the Speaker of the House recognizes the persons who would ask them, and thereby calls upon them to put them. Press conferences are like this too. They are usually chaired events, with the person in charge trying to direct, or at least influence the question-and-answer traffic. The person in charge will be unavailing if he lacks a fundamental power. It is the same power as the Speaker's. It is the power to recognize a questioner before his question can be asked. The power to recognize won't be worth having unless the person who wields it does so discriminately. Meetings will fail if irrelevant, rebarbative or damaging questions aren't stopped at their source. The requirement to be recognized is a constraint on freedom of speech, motivated by the institutional or political interests of those who called the meeting in the first place. No one who is good at this sort of thing can wield the power of recognition intelligently in the absence of some acquaintance with the people whose hands press for his attention. He should know whom his recognitions favour, but he should also know whom his non-recognitions exclude. Sometimes, in somewhat more formal settings, the person in charge will say to an expectant questioner, "Regrettably, sir, the chair does not see you." If Jones is the person to whom you say that you can't see him, it remains true that you see him and yet also that you don't, in the sense required for chairmanly recognition. "Not seeing", of course, is here a procedural idiom, with no iota of literal truth. But it makes a point of relevance to our own current preoccupations. It is that the unrecognizability of someone can be wholly at one with his concurrent presence in one's visual field.

The cognitive apparatus of the human reasoner is like this too. It serves to make

---

[32] We should also note that unrecognizability in this sense is not the effective unrecognizability or undecidability of metamathematics.

known facts unrecognizable, and consigns them to operational inertia. In environments that globalize it, provability is disabled for general use. It is an environment in which inconsistency cripples its theorems. This leaves two options to consider. In a system of crippled theorems, none is available for use in any context; or some are available for selective engagement only. On this view, unrecognizability serves as a kind of filtration device in which usable theorems are fit for selective premissory use, never mind that their negations are also theorems. Devices of this sort are a common part of the cognitive wherewithal of beings like us. Filtration devices help advance our cognitive agendas by screening out information that is irrelevant or too complex for timely processing. They are devices that cut-down large possibility spaces to smaller subsets of plausibility, and, as is the case here, they are the principal facilitators of premiss-selection in all contexts in which premisses are sought and useful to have. They serve a like purpose for what premisses are premisses of. For the purpose of conclusion-selection, they help to regulate the consequences that flow from those premisses. There are also awareness-filters. It is known that consciousness has a narrow bandwidth. It processes information very slowly. The processing rate of the five senses combined is in the neighbourhood of 11 million bits per second. For any of those seconds fewer than 40 bits make their way into consciousness. Consciousness is highly entropic, a thermodynamically expensive state for an information system to be in. And a good thing too! Filtration devices lend to the cognitive economy of the human animal efficiencies without which it would founder. For the most part, they operate automatically and out of sight of the mind's eye, and without the necessity or costs of engagement of our higher cognitive skills—another source of their efficiency. Unrecognizability is but one of these filtration devices and are no more mysterious or dubious than any of the others. When it is applied to systems of crippled theorems, the unrecognized ones are those the filter disables for agenda-advancing engagement.

8.4.4 Paraconsistency Without Tears

Paraconsistency, we saw, is usually taken as characterizing a family of logics in which *ex falso* is demolished, disarmed or evaded. In one large and prominent branch of that family with *ex falso* now out of the picture negation-inconsistencies are admissible, and yet are kept from causing further trouble by constraints on their usability as premisses.[33] All these logics brim with technical sophistication. They are logics that involve a lot of heavy lifting. There is a rule of thumb about this. The more a logic requires heavy lifting, the more it will need to call upon heavy equipment technologies. One reason

---

[33] The logics in question include the discussive logics of Jaśkowski, the (other) non-adjunctive logics of Rescher, Manor, and Rescher and Brandom, and the preservationist logics of Schotch, Jennings, and Brown, and also Brown and Priest, ("Chunk and permeate", *Journal of Philosophical Logic*, 33 (2004), 379-388). A particularly impressive technical means of effecting this disablement without undue collateral damage, is the nonmonotonic relation of *forcing*, which preserves something *likeable* in premisses short of truth, e.g. logical structure or coherence. See here Gillman Payette, "Preserving logical structure", in Schotch *et al.*, *On Preserving*, pages 105-143, and Jennings and Schotch, "The preservation of coherence", *Studia Logica*, 43 (1984), 89-106. See also Payette and Schotch, "Preserving what?", pages 85-104 of *On Preserving*; Nicolson and Brown, "Representation of forcing", pages 145-160, *ibid*; and Schotch, "Forcing and practical inference", pages 161-173, *ibid*.

why these logics retain their minority position is that the majority wants no truck at all with inconsistency. Another is that heavy equipment logics—those with lots of mathematical complexities in their formal arrangements—are hard to work with and not (some say) worth thinking up in the first place.[34]

If there were a systematic way of accounting for how Classicus arrived at all those set-theoretic truths in *An Introduction to Elementary Set Theory*, we could think of it as a premiss-constraining *unrecognizability logic*. An unrecognizability logic would offer itself as a friendly amendment to paraconsistent logic. It would express its friendliness in two ways. It would preserve the key feature of premiss-constraining logics by denying to theorems automatic eligibility for premissory engagement. It would also remove the technically heavy nuisance of keeping the inconsistencies they admit from the scourge of globalization. Both approaches generate utterly crucial premiss-restrictions. In discussive and non-adjunctive logics, we get premiss-selectivity with lots of hard work. In unrecognizability logics, we get premiss-selectivity without so much fuss. We might even say that with unrecognizability logics, we have paraconsistency without tears.

8.4.5 Can't-help-it Realism

I want now to return to an earlier point, with apologies again to those who dislike repetativeness as a matter of principle. The account I've given of the unrecognizability of absolute inconsistent pivots on two considerations. One is the empirical evidence of the cognitive utility of inconsistent theories. The other is the truth of *ex falso*. I mean by this its real truth, a truth that leaves its footprint on the world. Some will see this as naivety that strains against polite rejection. When I introduced my version of the Lewis-Langford proof, I said that I was "realistically minded" about *ex falso*. What I meant by that is that I take the realist stance to this question, as do we all in relation to all answers to all questions of interest. Like the rest of us, I take the realist stance as naturally as I breathe. Not taking it is not an option, not even for Gorgias. In advancing his sceptical claims, Gorgias really thought that he was "telling it like it is"; he was saying how he really thought things really go. This is not to disparage Gorgias, still less to mock him. The point that matters here is that Gorgias' situation is our situation too, never mind the differences in what each of us thinks is really so. Here I part from Schotch and Jennings. They recommend that we say to Lewis and my teacher Langford the following:

> "It's a bit silly to insist that we have a genuine implication (or a strict implication, if you must) between the propositions α and β only when one can deduce β from α. This is silly because we already have a representation for that in our logic, namely ⊢. So your strict implication relation is simply

---

[34] For more on this, readers could check my "Advice on the logic of argument", *Revista de Humanidades de Valparaiso*, I (2013), 7-29 and "The fragility of argument", in Fabio Paglieri, editor, *The Psychology of Argument*, London: College Publications, to appear in 2016.

8.4. HOW COME?

another word for provability—and why would there be need to rename provability?"[35]

The mind boggles, anyhow mine does. Does the provability version of my reworking of the Lewis-Langford demonstration fail? If so, where? If it does indeed by their lights fail there, I will say that in their ⊢ there is no recognizable presence of Lewis' and Langford's own provability, the notion they think adequately captured by their relation of strict implication.[36]

The realist stance conveys a considerable adaptive advantage. It is a primitively situated disposition for engagement with the world and ourselves. It is so deeply dug-in, and so necessary for experiential balance that the very idea of our breaking ourselves of this habit is laughable on its face. This makes us all can't-help-it realists, even those of who think that realism can't possibly be true.[37] In all our dismissals of the reality of things—whether phlogiston, Vulcan, Leibniz's infinitismals, the Abrahamic dieties, the biggest natural number, other minds, the past, the external world, and all the rest—we lodge our disavowals in how we really think things really go. We can no more still our can't-help-it realism than stop our blood flowing or our hair growing just by willing them to do so.

Hardly anyone thinks that the impulse to can't-help-it realism should be regretted across the board. Even big-box sceptics assume the stance with respect to the considerations that advance their scepticism. When someone takes the realist stance towards *ex falso*, those who oppose *ex falso* will take the realist stance towards the considerations they take to weigh against it. This raises a central question for in-house disagreements about when the stance should be taken and to what, as when logicians argue about whether a proof rule is actually valid. The question it raises is the question of question-begging, and with it the larger question of how fundamental disagreements of any kind can, in the absence of empirically available decision points, be managed at all without begging the question. It is the question posed by Frank Ramsey in *The Foundations of Mathematics* in 1932 of how a philosophical enquiry can be conducted without a perpetual *petitio principia*.

Ramsey's question is not our project here.[38] For the present, I would be more than happy if newcomers to sets checked out the rich instruction that awaits them in Classicus' little primer.

8.4.6 Inconsistency Robustness

I suppose that it can't be ruled out *à priori* that some readers would see the example of Classicus as doubtful assistance to the thesis I am advancing here. After all, *An Introduction to Elementary Set Theory* is a beginners' textbook isn't it, and don't we all

---

[35] Jennings and Schotch, "Paraconsistency: Who needs it?" in Schotch *et al.*, *On Preserving*, pages 17-31; section 2.2 "The Strange Case of C. I. Lewis", p. 21.
[36] Recall that at p. 259 of *Symbolic Logic*, Lewis and Langford insist that their proof holds for the ordinary meaning of implication, proof and inference. (Of course, insistence isn't truth-tracking either.)
[37] See my *Paradox and Paraconsistency*, xvi, 36, 124, 145-146, 149, 224-225, and 324.
[38] It is a principal part of the motivation for *Paradox and Paraconsistency*.

know that primers never get things right, not in philosophy anyhow? Had I the space, I'd say why any objection along such lines as this misses the point of my thesis just about as widely as possible. What I will do instead is offer a more grown-up example, an example that fairly brims with technical sophistication. This is the example of what Carl Hewitt calls inconsistency robustness, which "is performance of information systems with pervasively inconsistent information."[39] Although officially defined as a property of a system's performance, inconsistency robustness can also be seen as a property of the systems themselves when the following conditions are met:

- the system is large
- its inconsistency is pervasive, continual and indeed perpetual
- the system is practically useful
- its inconsistency is empirically discernible
- suppression or excision of their inconsistencies would defeat the system's practical utility, even if it were possible in principle.[40]

Inconsistency robust logic is a work in progress, calling upon what Hewitt capitalizes as Inconsistency Robust Direct Logic, which is "a minimal fix to Classical Logic without the rule of Classical Proof by Contradiction".[41] Further details are liberally available in the Hewitt-Woods volume. It is clear that Hewitt's is a paraconsistent logic. Less clear is whether it is the logic that underwrites the cognitive management of robustly inconsistent information systems. Writes Feferman,

> "So far as I know, it has not been determined whether such [inconsistency robust] logics account for "sustained ordinary reasoning", not only in everyday discourse but also in mathematics. And the sciences."[42]

Point to Feferman. No one in the IRL camp disagrees with that finding. They all agree that the question is still robustly open.[43]

My purpose in citing the Hewitt example is now easily explained. Inconsistency robust systems are no *rara avis*. They are utterly commonplace. The Five Eyes intelligence network is certainly one of them, and theories of climate modelling another. So too are the legal systems of every common law jurisdiction in recorded history. We must add

---

[39] Carl Hewitt, "Inconsistency robustness in foundations: Mathematics self proves its own consistency, and other matters", in Hewitt and Woods, *Inconsistency Robustness*, p. 104.

[40] Woods, "Inconsistency: Its present impacts and future prospects", in Hewitt and Woods, *op. cit.*, p. 158.

[41] More carefully, "because of pervasively inconsistent information, Inconsistency Robust Direct Logic theories must not have IGOR". IGOR acronymizes "inconsistency in, garbage out redux", in other words, *ex falso quodlibet*. See Hewitt, "Formalizing common sense reasoning for scalable inconsistency-robust information coordination using Direct Logic™ reasoning and the Actor model", Hewitt and Woods, p. 3.

[42] Solomon Feferman, "Axioms for determinateness and truth", *Review of Symbolic Logic*, 1 (2008) 204-217.

[43] Sixteen open questions can be found on pp. *xlix* and *l* in Hewitt and Woods, six more on p. *lxii*, and a further half dozen on the page that follows it.

8.4. HOW COME? 163

to the list "the community of professional mathematicians" (*p. xxviii*). The community of professional mathematicians is a large cognitive economy. Mathematicians aren't the sole possessors of one. All the rest of us, too operate from birth to life in communally interactive cognitive economies. Such economies are large and sufficiently stable and productive to grant us our survival and prosperity, and the means to create from time to time great civilizations. By any fair measures such economies qualify for the title of inconsistency robustness. They are large, pervasively and continually inconsistent and useful, and resistant to inconsistency-cleansing, which is practically impossible in any case. If what I say about IGOR is true, the human cognitive economy is also absolutely inconsistent. Its inconsistencies are not just pervasive and perpetual. They are global. The thesis that I've been proposing can now be seen in new and I think helpful light. If Hewitt and his IRL colleagues are right to say that we've somehow managed to stay productively afloat in the face of irremediable pervasive inconsistencies, how much more of a stretch could it be to do the same in the face of inconsistencies gone global?

Either way, something enables us to keep our cognitive balance. Establishment paraconsistentists put it all down to the consequence relation to operate the filtration-device that precludes detonation. I say that entailment's strictures are not where our cognitive stability reposes. I differ from the establishment in a supplementary way. The filters that keep us going and cognitively upright screen out something or other. But it's not absolute inconsistency. It is its inferential recognisability in everyday cognitive practice.[44]

8.4.7 A Proof Trichotomy

The time has arrived at which quickly to redeem a promissory note of section 1 concerning the distinction between the syntactic and semantic notions of proof. Doing so will motivate a further distinction that separates semantic proofs of the model-theoretic sort from semantic proofs that are natural language constructions, with respect to which the word "proof" carries its traditional mathematical and everyday meaning. Sound and complete logistic systems establish a tight extensional equivalence between its own syntactic proofs and its own model-theoretic ones. In model-theoretic environments "true" is a relational predicate. A wff of L is true on an interpretation I if and only if it has a model in I. It has a model in I just in case every countably infinite sequence from I's domain satisfies it. If A is an atomic truth in I, every such sequence pairs A to some undefined "semantic" object T, namely the value of I's valuation function on the atoms of I.

It is interesting to note that while "is true" is a predicate satisfied by meaningless strings A in I just when A has a model in I, it is not the predicate predicable of meaningful sentences of natural language. There are two good reasons for saying that in these two quite different contextual *milieu* the ordinary words "is true" bear entirely different meanings and entirely different application conditions. Here is a case in which ambiguity causes real trouble. As appended to a statement-expressing sentence of

---

[44] Mind you, there is an important difference between Classicus and Hewitt. Classicus' take on inconsistency is classical. Hewitt is as classical as he can afford to be; but assuredly not so about inconsistency robustness.

English, "is true" is a content-reitorator whereas, when applied to a meaningless string of a meaningless language constrained by I, there is no content to reiterate.[45] For any A of I, all statements of the form ⌜A is T⌝ are ill-formed. The point of raising these matters here bears directly on the relevance of sound and complete logistic systems for the issues raised by the English version of the Lewis-Langford proof. Consider an example. Let LS be a logistic system over an artificial "language" L for which its "logical truths" and its "theorems" stand to one another in a tight isomorphism that preserves properties of interest. In that case, we have it that ⊨ A if and only if we also have it that ⊢ A. There is a rude way of asking a good question: What bearing do these one-to-one correspondences have upon whether my reworking of the Lewis-Langford demonstration is cogent. In the absence of a formal representability proof that matches LS's logical truths to those expressible in English, and LS's theorems to those that are cobbled together in English, the answer must surely be "None to speak of."

It bears repeating that in natural language contexts, the English word "true" is also a predicate, not of two places but only of one. What it predicates is the property of being true, and what it predicates it of is not a wff of a system like L, but rather of an English declarative sentence (or sentences sometimes—let's not forget "Everything Nixon says is true"). When philosophers ask about truth-trackers they aren't asking about tracking a model in some I. They are asking about what, since Tarski, they refer to as the intuitive property truth. We've already seen that there are perfectly made logistic systems that don't hook up with any truth answering to their theorems, whether syntactically or model-theoretically wrought. No direct derivation is a proof in the mathematician's intuitive sense under each of its contained lines is true in the intuitive sense. Direct proofs in the model-theoretic sense carry a heavier burden. Each contained line must have a model in every set-theoretic structure serving as the interpretations I of L. Model-theoretic proofs are, as we might say, L-truth-tracking but they aren't, just so, truth-tracking or truth-preserving either. To be sure, sometimes theory-based proofs—of a theory of sets, for example—pair up with intuitive proofs. When this happens, we *might* say that what the conclusion of the theory's proofs are true *of* is sets, but we cannot always say with accuracy that they themselves *express* those truths.

This helps in understanding what Classicus had a nose for in *An Introduction to Elementary Set Theory*. In certain cases, what he told his readers about sets was false, even when an axiom of the naïve theory of sets. In other cases, what he told them about sets he proved to be true. Classicus' nose enabled him to filter from the teeming inconsistencies of the theory the proofs that chanced to be truth-tracking. Thus did he convey to his readers enough knowledge of sets to enable them to appreciate the necessity of having them in the foundations of arithmetic. The set theoretic proofs that Classicus' filtration device left behind had no recognizable presence in this transmission of knowledge, never mind the primer's own absolute inconsistency. That's quite some nose that Classicus had. It turns out that to some extent or another the rest of us have it

---

[45]Logical truth for formal "languages" reflect this same alienation from logical truths of English. A is logically true in a logistic system L just when every countably infinite sequence of the objects of the domain satisfies it in all interpretations. Another reason, we might suppose, not to admit the English predicate "is true" into the model theory of I is Tarski's demonstration of its inconsistency. (I have my doubts about this; but there is no space for them here.)

8.4. HOW COME?

too. In theories of inconsistency-management it is a fact that deserves the prominence due it, not least because it spares us the nuisance and embarrassment of having to go ballistic over *ex falso*.

Kudos: The turbulence Peter Schotch began to stir up in Waterloo in 1975 gathered early momentum and, in partnership with Ray Jennings, evolved into Canada's most intellectually powerful contribution to modern logic to date, and not by any means easily bettered. Unlike other developments in Canadian logic—notably the restoration of fallacy theory to the project of logic and the origination of literary semantics as a stand-alone approach to the non-existent objects of fiction—the preservationist pioneers have nourished a stable and productive successorship: Bryson Brown was, in short order, first to join up, with Blaine d'Entremont and David Johnson arriving soon after. In due course, came Dorion Nicholson, Darko Sarenac and Gill Payette and others.

I am delighted to dedicate this essay to the volume's distinguished honouree.

Acknowledgements: An earlier version of this essay was delivered as a keynote lecture to the Square of Opposition conference at Rome's Lateran Pontifical University in June 2014. For welcome and helpful comments in Rome or in follow-up correspondence, I thank Lorenzo Magnani, Dale Jacquette, John N. Martin, Jonathan Westphal, Rusty Jones, Jean-Yves Béziau, Dov Gabbay, Daniel Clausén, and others whose names I lack or haven't managed to remember. Apologies. A follow-up paper, under the title "How I stopped worrying, and learned to live with inconsistency", was given as the keynote lecture to the Western Canadian Philosophical Association, at UBC in October 2014, by kind invitation of its organizers Sylvia Berryman and Carrie Jenkins. For helpful comments or stimulating resistance, I am grateful to Roberta Ballarin, Adam Morton, Mark Migotti, and Jonathan Ichikawa. A version of that version under the present title was presented in May 2016 to the Philosophy Colliguium of the University of Konstanz, by kind invitation of Wolfgang Spohn. My appreciation to David Atkinson, Jeanne Peijenburg and Wolfgang Spohn for testing questions, and to several others whose names I lack. Apologies again. I especially welcome Gillman Payette's shrewd instruction on critical points, and thank him for his patient editorship of this volume. For technical and administrative support I am indebted, as in all things, to Carol Woods.

Chapter 9

From Jónsson and Tarski to Schotch and Jennings

Alasdair Urquhart[1]

9.1 Introduction

The work of R.E. Jennings and P.K. Schotch on weakly aggregative modal logics and preservationism arose originally from concerns about the adequacy of the standard modal semantics in the context of deontic logic. When the modal operators $\Box$ and $\Diamond$ are interpreted in terms of deontic concepts, then the usual modal semantics validates principles that do not seem to be in accord with our intuitions about "ought" and "can." This led Jennings and Schotch to logics in which the modal aggregation principle

$$K: \quad \Box\alpha \wedge \Box\beta \implies \Box(\alpha \wedge \beta)$$

is replaced by weaker versions such as the scheme of "n-ary aggregation":

$$K_n: \quad \Box A_0 \wedge \cdots \wedge \Box A_n \to \Box[\bigvee_{0 \leq i < j \leq n} A_i \wedge A_j].$$

For more details on these developments, the reader can consult the excellent history Brown (2007) of preservationism.

There is another point of view, however, from which we can approach weakly aggregative modal logics, namely that of algebraic logic. In fact, this theme emerged early in the developments leading to the preservationist project. Bryson Brown reports in his historical survey:

> In the summer of 1976, Schotch recalled work he had done in the algebra of modal logic while at the University of Waterloo. Denis Higgs, an algebraist there, had introduced Schotch to Jónsson and Tarski (1951),

---

[1] Department of Computer Science, University of Toronto, urquhart@cs.toronto.edu.

"Boolean Algebra with Operators", which included consideration of n-ary frames: A ternary frame corresponds to a binary operator in Tarski and Jónsson's approach, and in general, an n-ary frame corresponds to an $n-1$-ary operator. Higgs had encouraged Schotch to work on modal logic from this point of view (Brown, 2007, p. 102).

The present paper contains some developments along the lines originally suggested by Denis Higgs. It is a sequel to an earlier article Urquhart (2009) in the same area. There is some overlap with that paper, but here we attempt to set the developments in a more general historical context, and in addition, we have included some unpublished material that was omitted from the earlier piece.

Before beginning the main development, it is convenient to introduce notation for sequences. If S is a non-empty set, we use vector notation $\vec{a}, \vec{b}, \ldots$ to refer to sequences of elements from S. We use $|\vec{a}|$ for the length of the sequence $\vec{a}$; the sequence of length $k > 0$ in which all of the entries are the element a is written a^k. If $\vec{k}$ is a sequence of integers, then we write $\sum \vec{k}$ for the sum of the elements in the sequence.

We write $x \in \vec{y}$ if x is a member of the sequence $\vec{y}$. If $\vec{x}$ is a sequence of elements from a set S and $A \subseteq S$, then we write $\vec{x} \subseteq A$ if every element in the sequence $\vec{x}$ is in A. If $\vec{X}$ is a sequence of subsets of a fixed set S, then we use the notation $\bigcap \vec{X}$ for the intersection of all the sets in the sequence $\vec{X}$. If $\vec{x}$ and $\vec{y}$ are sequences of elements from a set with an ordering relation $\leq$ defined on it, and $|\vec{x}| \leq |\vec{y}|$, then we write $\vec{x} \leq \vec{y}$ if $x_i \leq y_i$ for all i, where $1 \leq i \leq |\vec{x}|$. For $\vec{x}$ and $\vec{y}$ sequences, we write $\vec{x}^\frown \vec{y}$ for the sequence obtained by concatenating $\vec{x}$ and $\vec{y}$.

9.2 Boolean Algebras with Operators

In the early 1950s, Bjarni Jónsson and his thesis advisor Alfred Tarski published two papers Jónsson and Tarski (1951, 1952) containing deep results on Boolean algebras with operators. The first of these articles gives a general representation theorem for such algebras, while the second contains applications of the general theory to relation algebras. We shall only be concerned with the first paper here, though the second is also a remarkable piece of work. These two papers already contain many results that were rediscovered later by authors working on nonclassical logics, and in particular completeness theorems for propositional modal logics. As we shall see below, they even anticipated some of the work of Schotch and Jennings, though in a rather hidden way.

The following definition is from (Jónsson and Tarski, 1951, Definition 1.1); for convenience, we have added the abbreviation "multimodal" to their list of three properties.

Definition 9.2.1. *Let $\mathfrak{B} = \langle B, \wedge, \vee, \neg, 0, 1 \rangle$ be a Boolean algebra, and ∇ an n-place operation on $\mathfrak{B}$. Then ∇ is called*

1. normal *if $\nabla(\vec{a}, 0, \vec{b}) = 0$, where $\vec{a}^\frown 0^\frown \vec{b} \in B^n$;*

2. monotone *if $\nabla(\vec{a}) \leq \nabla(\vec{b})$ whenever $\vec{a} \leq \vec{b}$, for $\vec{a}, \vec{b} \in B^n$;*

9.2. BOOLEAN ALGEBRAS WITH OPERATORS

3. additive *if* $\nabla(\vec{a}, b \vee c, \vec{d}) = \nabla(\vec{a}, b, \vec{d}) \vee \nabla(\vec{a}, c, \vec{d})$, *where* $\vec{a}\hat{\ }\vec{d} \in B^{n-1}$;

4. multimodal *if it is normal, monotone and additive.*

The familiar relational semantics for modal logic, expounded by Saul Kripke in the 1960s, analyses one-place modal operators in terms of a binary accessibility relation. This idea was anticipated by Jónsson and Tarski, and even in a more general form, since they analyse n-place operators in terms of $n + 1$-place relations. The next definition is Definition 3.2 from (Jónsson and Tarski, 1951, p. 929).

Definition 9.2.2. *Let U be a set, R an $n+1$-place relation on U, and $P(U)$ the Boolean algebra of all subsets of U. If $\vec{A} \in P(U)^n$, then we define an n-place operation ∇_R on $P(U)$ by*

$$\nabla_R(A_1, \ldots, A_n) = \{x \mid \exists \vec{y} \in U^n (Rx\vec{y} \wedge \forall i(y_i \in A_i))\}.$$

The special case of the above definition where R is a binary relation corresponds exactly to Kripke's analysis of possibility, where we think of U as a set of "possible worlds." The main representation theorem of (Jónsson and Tarski, 1951, Theorem 3.10) shows that every multimodal operator on a Boolean algebra can be represented in the manner of Definition 9.2.2. Let us define the algebra $\langle P(U), \nabla_R \rangle$ defined above to be the *complex algebra* of the relation R.

Theorem 9.2.3.   *1. If U is a set, R an $n + 1$-place relation on U, and $P(U)$ the Boolean algebra of all subsets of U, then ∇_R is a multimodal operator on $P(U)$;*

2. *Let $\mathfrak{A} = \langle B, \wedge, \vee, \neg, 0, 1, \nabla \rangle$ be an algebra where $\mathfrak{B} = \langle B, \wedge, \vee, \neg, 0, 1 \rangle$ is a Boolean algebra and ∇ is a multimodal n-place operator on $\mathfrak{B}$. Then $\mathfrak{A}$ is isomorphic to a subalgebra of the complex algebra of an $n + 1$-place relation.*

In §9.4, we state a more general representation theorem from which this result of Jónsson and Tarski can be obtained as a corollary.

Theorem 9.2.3, restricted to the case of a one-place operator, is essentially the same as Kripke's completeness theorem for the logic **K**. What is more, some of the well known correspondence theorems relating properties of modal operators to the properties of accessibility relations also appear as Theorem 3.5 of their first paper Jónsson and Tarski (1951). Kripke himself only became aware of this anticipation after proving his completeness theorems independently; in a lengthy footnote (Kripke, 1963, p. 69) to his famous paper on normal modal logics, listing anticipations of his semantical analysis, he remarks:

> The most surprising anticipation of the present theory, discovered as this paper was almost completed, is the algebraic analogue in Jónsson and Tarski (1951). Independently and in ignorance of Jónsson and Tarski (1951) (though of course much later), the present writer derived its main theorem by an algebraic analogue of his semantical methods; the proof will appear elsewhere.

We may wonder why the work of Jónsson and Tarski escaped the attention of logicians for over a decade. Part of the reason, no doubt, was that they published their paper in a mathematical journal not usually read by logicians. Furthermore, they fail to mention the connections with modal logic, and their paper lacks the colourful vocabulary of possible worlds invented by Kripke. We might recall the title of a paper Fleischer (1984) by the mathematician Isidore Fleischer: *"Kripke semantics" = algebra + poetry*. Fleischer begins his article with these remarks:

> The tenor of the argument should be sufficiently clear from the title. The thesis is basically that the algebra is already quite well known, while the additional poetry only obscures and unnecessarily complicates the essential aspects of the situation as they will be revealed to those incisive enough to forego the imaginative sugarcoating.
>
> The relevant algebraic analysis often enough appears in the literature (more often for the modal logics, where the facts are closer to the surface, than for the intuitionistic) still enmeshed in the symbolism of the initially presented formal systems and doused more or less liberally with "possible world semantics" sauce ...

9.3 Diagonalizing Multimodal Operators

In the remainder of the paper, we shall work in the context of bounded distributive lattices with operators, since the complement operation is not needed for what follows. This makes the results slightly more general, since those for Boolean algebras follow as corollaries. Thus the algebras on which we define operators are of the form $\mathfrak{D} = \langle D, \wedge, \vee, 0, 1 \rangle$, where $\mathfrak{D}$ is a bounded distributive lattice. Definition 9.2.1 applies in this context, since the complement operator is not used in it. We define a *multimodal algebra* to be of the form $\mathfrak{A} = \langle D, \wedge, \vee, 0, 1, \nabla \rangle$, where $\mathfrak{D} = \langle D, \wedge, \vee, 0, 1 \rangle$ is a bounded distributive lattice and ∇ a multimodal operator on $\mathfrak{D}$.

If $f(x_1, \ldots, x_n)$ is an n-place operator on an algebra, then we define the *diagonalization* of f to be the one-place operator $f(x_1, \ldots, x_1)$; the terminology is that of Leung and Jennings (2009). If ∇ is a multimodal operator on a Boolean algebra, then its diagonalization may not be additive. For example, if U is the set $\{1,2,3\}$, and $R = \{\langle 1,2,3\rangle\}$, then the operator defined on $P(U)$ by the construction of Theorem 9.2.3 is multimodal, but its diagonalization is not, since $\nabla_R(\{2,3\},\{2,3\}) = \{1\}$, but $\nabla_R(\{2\},\{2\}) = \nabla_R(\{3\},\{3\}) = \varnothing$. This is of course just an algebraic expression of the failure of the modal aggregation principle, familiar from the work of Schotch and Jennings.

Jennings and Schotch discovered that the diagonalization of a multimodal n-place operator satisfies a weaker form of aggregation, the scheme $\mathbf{K}_n$ of "n-ary aggregation":

$$\Box A_0 \wedge \cdots \wedge \Box A_n \to \Box [\bigvee_{0 \leq i < j \leq n} A_i \wedge A_j],$$

9.3. DIAGONALIZING MULTIMODAL OPERATORS

or in the dual form that we are using here:

$$\Diamond[\bigwedge_{0 \le i < j \le n} A_i \vee A_j] \to \Diamond A_0 \vee \cdots \vee \Diamond A_n.$$

It is possible to generalize the diagonalization construction further. When we diagonalize an operator, we identify all of its argument places; it is possible, though, to identify some of the arguments but not others. For example, if $f(x_1, x_2, x_3, x_4, x_5)$ is a five-place operator then we can form a three-place operator $g(x_1, x_2, x_3) = f(x_1, x_1, x_2, x_3, x_3)$. The next definition introduces a notation for such constructions.

Definition 9.3.1. *Let X be a non-empty set, and ∇ a function from X^n to X. If n is a positive integer, and $\vec{k} = \langle k_1, \ldots, k_j \rangle$ a vector of positive integers such that $\sum \vec{k} = n$, we define the $\vec{k}$-collapse of ∇ to be the operator $\Diamond$ defined by*

$$\Diamond(a_1, \ldots, a_j) = \nabla(a_1^{k_1} \frown a_2^{k_2} \ldots \frown a_j^{k_j}).$$

As an example, the operator g defined in the previous paragraph is the $\langle 2, 1, 2 \rangle$-collapse of the operator f. If f is an n-place operator, then f itself is the 1^n-collapse of f, while the diagonalization of f is the $\langle n \rangle$-collapse of f.

Let $\Diamond$ be an n-place operation defined on a bounded distributive lattice L. For $k > 0$, we say that $\Diamond$ is k-additive in the ith place if it satisfies the conditions:

$$\Diamond[\vec{a}, \bigwedge_{0 \le h < j \le k}(b_h \vee b_j), \vec{c}] \le \bigvee_{0 \le j \le k} \Diamond[\vec{a}, b_j, \vec{c}],$$

for $\vec{a} \in L^{i-1}$, $\vec{c} \in L^{n-i}$ and $b_0, \ldots, b_k \in L$. We define an operator on a bounded distributive lattice to be *weakly additive* if in all of its argument places it is k-additive for some $k > 0$; it is *weakly multimodal* if it is normal, monotone and weakly multimodal. A multimodal operator in the sense of Jónsson and Tarski (1951, 1952) is a weakly multimodal operator that is 1-additive in all of its argument places.

We say that an n-place weakly multimodal operator $\Diamond$ is of type $\vec{k}$ if $\vec{k}$ is a vector of positive integers of length n so that $\Diamond$ is k_i-additive in its ith place. For example, if $\Diamond(a, b, c)$ is 1-additive in its first place, 3-additive in its second place and 2-additive in its third, then it is of type $\langle 1, 3, 2 \rangle$. Thus, an n-place multimodal operator is of type 1^n, while its diagonalization is a one-place operator of type $\langle n \rangle$. We define a *weakly multimodal algebra of type $\vec{k}$* to be a bounded distributive lattice L together with a weakly multimodal operator of type $\vec{k}$ defined on L.

Lemma 9.3.2. *Let ∇ be an n-place additive operator on a distributive lattice, and $\vec{k}$ a vector of positive integers, with $\sum \vec{k} = n$. If $\Diamond = \nabla^{\vec{k}}$ is the $\vec{k}$-collapse of ∇, then $\Diamond$ is a weakly additive operator of type $\vec{k}$.*

Proof. Let $k = k_i$ be an element of the sequence $\vec{k}$. We wish to show that the operator $\Diamond$ is k-additive in the ith place. This is proved as follows:

$$\Diamond[\vec{a}, \bigwedge_{0 \le h < j \le k}(b_h \vee b_j), \vec{c}\,] = \nabla(\vec{d}, [\bigwedge_{0 \le h < j \le k}(b_h \vee b_j)]^k, \vec{e}\,) \quad (9.1)$$

$$= \nabla(\vec{d}, [\bigvee_{\substack{X \subseteq \{0,\ldots,k\} \\ |X|=k}} \bigwedge_{h \in X} b_h]^k, \vec{e}\,) \quad (9.2)$$

$$= \bigvee_{\substack{X_i \subseteq \{0,\ldots,k\} \\ |X_i|=k}} \nabla(\vec{d}, \bigwedge_{h \in X_1} b_h, \ldots, \bigwedge_{h \in X_k} b_h, \vec{e}\,) \quad (9.3)$$

$$\le \bigvee_{0 \le h \le k} \nabla(\vec{d}, (b_h)^k, \vec{e}\,) \quad (9.4)$$

$$= \bigvee_{0 \le h \le k} \Diamond(\vec{a}, b_h, \vec{c}\,), \quad (9.5)$$

where $\vec{d}$ and $\vec{e}$ are the result of replacing a_j and c_j in the vectors $\vec{a}$ and $\vec{c}$ by the corresponding vectors $(a_j)^{k_j}$ etc. In the proof of k-additivity above, the equality (2) follows by an identity valid in any distributive lattice; it can be seen to hold by virtue of the following fact: if $X \subseteq \{0, \ldots, k\}$ and $X \cap Y \ne \emptyset$ for all 2-element subsets Y of $\{0, \ldots, k\}$, then $|X| \ge k$. The equality (3) follows by the additivity of ∇. The inequality (4) follows by the fact that if $X_1, \ldots, X_k$ is a family of subsets of $\{0, \ldots, k\}$, with $|X_i| = k$, then $X_1 \cap \cdots \cap X_k \ne \emptyset$ (a manifestation of the pigeonhole principle). $\square$

9.4 Representation and Duality for Objects

In this section, we place Jónsson and Tarski's basic representation theorem for Boolean algebras with operators, Theorem 9.2.3, in a more general context. Their representation theorem can be proved in a stronger form where each such algebra corresponds to an ordered topological space from which it can be recovered exactly. In Urquhart (2009), this topological duality for algebras was stated and proved. An earlier, unpublished paper of which Urquhart (2009) is an edited version, contained a full duality theorem showing that the category of weakly multimodal algebras is dual to a certain category of ordered topological spaces. We shall prove this result in §9.5.

The duality for weakly multimodal operators described in this section is piggybacked on Hilary Priestley's duality for bounded distributive lattices. We begin by giving the main results of this duality theory without proofs. For detailed discussion of the theory, the reader is referred to the articles Priestley (1970, 1984), Davey and Duffus (1982) and to the textbook Davey and Priestley (1990).

An *ordered topological space* is a topological space with a partial order relation defined on it. A subset E of a partially ordered set is *increasing* if $x \in E$, $x \le y$ imply $y \in E$. A map f between partially ordered sets is *increasing* if $x \le y$ implies $f(x) \le f(y)$. A map between two ordered spaces is said to be an *order-homeomorphism* if it is both a homeomorphism and an isomorphism with respect to the orderings on the spaces. A *Priestley space* is an ordered topological space $\mathcal{S}$ that is compact and totally order-disconnected, that is, for points x, y of $\mathcal{S}$, if $x \not\le y$ then there is a clopen increasing set U such that $x \in U, y \notin U$.

9.4. REPRESENTATION AND DUALITY FOR OBJECTS

If L is a bounded distributive lattice, then the *dual space* of L, $\mathcal{S}(L)$, is the ordered topological space in which the set of points S is the family of all prime filters of L, ordered by containment, and the topology is defined by taking as a sub-base the family of all sets $\{x \in S : a \in x\}$ and $\{x \in S : a \notin x\}$ for $a \in L$. Conversely, if $\mathcal{S}$ is a Priestley space then the dual lattice of $\mathcal{S}$, $L(\mathcal{S})$, is the set of all clopen increasing sets of $\mathcal{S}$, with the lattice operations of set intersection and union.

Theorem 9.4.1. *1. If L is a bounded distributive lattice, then $\mathcal{S}(L)$ is a Priestley space, and L is isomorphic to $L(\mathcal{S}(L))$ under the mapping*

$$\eta(a) = \{x \in \mathcal{S}(L) : a \in x\};$$

2. If $\mathcal{S}$ is a Priestley space, and $L(\mathcal{S})$ its dual lattice, $\mathcal{S}$ is order-homeomorphic to $\mathcal{S}(L(\mathcal{S}))$ under the mapping

$$\theta(x) = \{B \in L(\mathcal{S}) : x \in B\}.$$

We obtain the duality theory for weakly multimodal operators by building on Priestley's duality theory. First, we need a generalization of Definition 9.2.2:

Definition 9.4.2. *Let U be a set, R an $n + 1$-place relation on U, and $P(U)$ the Boolean algebra of all subsets of U. We define an n-place operation $\Diamond_R$ on $\mathcal{P}(U)$ by the following definition:*

$$x \in \Diamond_R(A_1, \ldots, A_n) \Leftrightarrow \exists \vec{y_1}, \ldots, \vec{y_n}[Rx\vec{y_1}, \ldots, \vec{y_n} \wedge \forall i[1 \leq i \leq n \Rightarrow \vec{y_i} \subseteq A_i]],$$

where for each i, $1 \leq i \leq n$, $\vec{y_i}$ is a vector of length k_i.

If $\vec{k}$ is a sequence of positive integers, we define a *relational space of type $\vec{k}$* to be a structure $\langle \mathcal{S}, R \rangle$, where $\mathcal{S}$ is a Priestley space, and R is a $(\sum \vec{k} + 1)$-place relation on S, satisfying the following conditions:

1. If $A_1, \ldots, A_n \in L(\mathcal{S})$, then $\Diamond_R(A_1, \ldots, A_n)$ is clopen;

2. If $Rx\vec{y}$ and $x \leq z$ then $Rz\vec{y}$;

3. If for all i, $1 \leq i \leq n$, $|y_i| = k_i$ and $\neg Rx\vec{y_1}, \ldots, \vec{y_n}$ then:

$$(\exists A_1, \ldots, A_n \in L(\mathcal{S}))[\forall i(1 \leq i \leq n \Rightarrow \vec{y_i} \subseteq A_i) \wedge x \notin \Diamond_R(A_1, \ldots, A_n)].$$

If $\mathcal{R} = \langle \mathcal{S}, R \rangle$ is a relational space of type $\vec{k}$, then the *dual algebra* of $\mathcal{R}$ is the algebra $A(\mathcal{R})$ defined on the lattice $L(\mathcal{S})$ by adding the operation $\Diamond_R$. If $\langle L, \Diamond \rangle$ is a weakly multimodal algebra, where $\Diamond$ is an n-place operator, and each $\vec{y_1}, \ldots, \vec{y_n}$ is a sequence of filters in L, then the relation $R_\Diamond$ is defined by:

$$R_\Diamond x \vec{y_1}, \ldots, \vec{y_n} \Leftrightarrow \forall \vec{a} \in L^n [\, \forall i(a_i \in \bigcap \vec{y_i}) \Rightarrow \Diamond(a_1, \ldots, a_n) \in x \,].$$

If $A = \langle L, \Diamond \rangle$ is a weakly multimodal algebra of type $\vec{k}$, then the dual space of A, $\mathcal{R}(A)$, is defined by adding to the Priestley space of L the relation R defined by:

$$Rx\vec{y}_1, \ldots, \vec{y}_n \Leftrightarrow R_\Diamond x\vec{y}_1, \ldots, \vec{y}_n \wedge \forall i \, [1 \leq i \leq n \Rightarrow |\vec{y}_i| = k_i],$$

where $n = |\vec{k}|$.

The next lemma is the basic result used in proving the representation and duality theorems that follow.

Lemma 9.4.3. *Let $\langle L, \Diamond \rangle$ be a weakly multimodal algebra of type $\vec{k}$, where $|\vec{k}| = n$. Then for $a \in L$, $x \in \mathcal{S}(L)$,*

$$\Diamond \vec{a} \in x \Rightarrow \exists \vec{w}_1, \ldots, \vec{w}_n [\, Rx\vec{w}_1, \ldots, \vec{w}_n \wedge \forall i [\, 1 \leq i \leq n \Rightarrow a_i \in \bigcap \vec{w}_i \,] \,].$$

My paper Urquhart (2009) has a detailed proof of Lemma 9.4.3, as well as a proof of the following representation theorem from this lemma.

Theorem 9.4.4. *1. If A is a weakly multimodal algebra of type $\vec{k}$, then $\mathcal{R}(A)$ is a relational space of type $\vec{k}$ and A is isomorphic to $A(\mathcal{R}(A))$ under the mapping*

$$\eta(a) = \{x \in \mathcal{R}(A) : a \in x\};$$

2. If $\mathcal{R}$ is a relational space of type $\vec{k}$, and $A(\mathcal{R})$ its dual algebra, $\mathcal{R}$ is order-homeomorphic to $\mathcal{R}(A(\mathcal{R}))$ under the mapping

$$\theta(x) = \{B \in A(\mathcal{R}) : x \in B\}.$$

Theorem 2.3 of Jónsson and Tarski (1951) follows as a corollary of the first part of Theorem 9.4.4, by specializing to weakly multimodal algebras of type 1^n.

9.5 Duality for Maps

A significant trend in modern mathematics is that of emphasizing not only classes of mathematical objects, but the structure-preserving maps between objects in these classes. A family of objects, together with the corresponding family of maps, constitutes a *category*. We refer the reader in need of background to one of the many good introductory texts on category theory, for example, the well known book Mac Lane (1971).

We can often gain insight into a category **C** of algebras by proving a *duality theorem*. This amounts in abstract terms to describing the category that is dual to **C**, that is, the category where we reverse all the maps in **C**. Here are two simple examples of this process. The dual of the category of finite Boolean algebras is the category of finite sets with the maps the functions between finite sets. The dual of the category of finite distributive lattices is the category of finite partially ordered sets, with the maps the increasing functions between them. Theorem 9.5.1 below is a generalization of these two elementary examples.

9.5. DUALITY FOR MAPS

In the category BDL of bounded distributive lattices, the maps are the lattice homomorphisms that preserve the lattice bounds as well as the lattice operations. In the category PS of Priestley spaces, the maps are continuous increasing maps between spaces. If L_1, L_2 are bounded distributive lattices and ϕ a lattice homomorphism from L_1 to L_2, the dual map of ϕ, $\mathcal{S}(\phi)$, is the map from $\mathcal{S}(L_2)$ to $\mathcal{S}(L_1)$ defined by: $(\mathcal{S}(\phi))(\nabla) = \phi^{-1}(\nabla)$ for ∇ a prime filter in L_2. In the opposite direction, if $\mathcal{S}_1, \mathcal{S}_2$ are Priestley spaces and ψ a continuous increasing map from $\mathcal{S}_1$ to $\mathcal{S}_2$, then the dual map of ψ, $L(\psi)$, is the map from $L(\mathcal{S}_2)$ to $L(\mathcal{S}_1)$ defined by $(L(\psi))(A) = \psi^{-1}(A)$ for $A \in L(\mathcal{S}_2)$.

Theorem 9.4.1 gives the duality for objects; the following theorem extends it to a full duality for maps, showing that the category of Priestley spaces is dual to the category of bounded distributive lattices.

Theorem 9.5.1. *1. If $\mathcal{S}_1$ and $\mathcal{S}_2$ are Priestley spaces, then the map $L : \psi \mapsto L(\psi)$ is a bijection between PS maps from $\mathcal{S}_1$ to $\mathcal{S}_2$ and BDL maps from $L(\mathcal{S}_2)$ to $L(\mathcal{S}_1)$;*

2. If L_1, L_2 are bounded distributive lattices, then the map $\mathcal{S} : \phi \mapsto \mathcal{S}(\phi)$ is a bijection between BDL maps from L_1 to L_2 and PS maps from $\mathcal{S}_2$ to $\mathcal{S}_1$.

We construct the duality for maps for varieties of weakly multimodal algebras by piggybacking it on the duality of Theorem 9.5.1. We shall confine our attention to the variety of weakly multimodal algebras of type $\vec{k}$. Let $\mathcal{R}_1$ and $\mathcal{R}_2$ be two relational spaces of type $\vec{k}$, and ψ a continuous increasing map from $\mathcal{R}_1$ to $\mathcal{R}_2$. Then ψ is a *relational map* from $\mathcal{R}_1$ to $\mathcal{R}_2$ if the following two conditions are equivalent, for all $B_1, \ldots, B_n \in L(\mathcal{R}_2)$:

1. $\exists \vec{y}_1 \ldots \vec{y}_n [R_1 x \vec{y}_1, \ldots, \vec{y}_n \land \forall i (1 \leq i \leq n \Rightarrow \psi(\vec{y}_i) \subseteq B_i)]$;

2. $\exists \vec{z}_1 \ldots \vec{z}_n [R_2 \psi(x) \vec{z}_1, \ldots, \vec{z}_n \land \forall i (1 \leq i \leq n \Rightarrow \vec{z}_i \subseteq B_i)]$.

In the next theorem, we prove a full duality between the category of weakly multimodal algebras of type $\vec{k}$ and relational spaces of type $\vec{k}$.

Theorem 9.5.2. *1. If $\mathcal{R}_1$ and $\mathcal{R}_2$ are relational spaces of type $\vec{k}$, then the map $L : \psi \mapsto L(\psi)$ is a bijection between relational maps from $\mathcal{R}_1$ to $\mathcal{R}_2$ and homomorphisms from $A(\mathcal{R}_2)$ to $A(\mathcal{R}_1)$;*

2. If A_1, A_2 are weakly multimodal algebras of type $\vec{k}$, then the map $\mathcal{S} : \phi \mapsto \mathcal{S}(\phi)$ is a bijection between homomorphisms from A_1 to A_2 and relational maps from $\mathcal{R}(A_2)$ to $\mathcal{R}(A_1)$.

Proof. Let $\mathcal{R}_1$ and $\mathcal{R}_2$ be relational spaces of type $\vec{k}$, and ψ a relational map from $\mathcal{R}_1$ to $\mathcal{R}_2$. Furthermore, let the dual map $L(\psi) = \phi$ be defined by $\phi(B) = \psi^{-1}(B)$, for $B \in L(\mathcal{R}_2)$. If $B_1, \ldots, B_n \in L(\mathcal{R}_2)$, $x \in \mathcal{R}_2$, the condition $x \in \Diamond_1(\phi(B_1), \ldots, \phi(B_n))$ is equivalent to the first condition in the definition of a relational map. Similarly, the condition $x \in \phi(\Diamond_2(B_1, \ldots, B_n))$ is equivalent to the second condition in the definition of a relational map, showing that $L(\psi) = \phi$ is a homomorphism from $A(\mathcal{R}_2)$ to $A(\mathcal{R}_1)$.

Conversely, let A_1, A_2 be weakly multimodal algebras of type $\vec{k}$, and ϕ a homomorphism from A_1 to A_2. In addition, let $\psi = \mathcal{S}(\phi)$ be the dual map defined by $\psi(\nabla) = \phi^{-1}(\nabla)$. We need to show that the two conditions in the definition of a relational map are equivalent, for the function ψ just defined. Now if $\mathcal{R}_2$ is the dual space of A_1, and $B_1, \ldots, B_n \in L(\mathcal{R}_2)$, then by Theorem 9.4.4, there are $b_1, \ldots, b_n \in A_2$ so that $B_1 = \eta(b_1), \ldots, B_n = \eta(b_n)$. Consequently, the first condition can be rewritten as:

$$\exists \vec{y}_1 \ldots \vec{y}_n [R_1 x \vec{y}_1, \ldots, \vec{y}_n \wedge \forall i (1 \leq i \leq n \Rightarrow \phi(b_i) \in \bigcap \vec{y}_i)],$$

which by Lemma 9.4.3 is equivalent to $\Diamond_1(\phi(b_1), \ldots, \phi(b_n)) \in x$. Similarly, the second condition is equivalent to $\phi(\Diamond_2(b_1, \ldots, b_n)) \in x$, showing that the two conditions are equivalent. □

9.6 Composition of Multimodal Operators

We noted above in §9.2 that the work of Jónsson and Tarski on Boolean algebras with operators anticipated a good deal of the discoveries of the 1960s in the area of modal logic. In the present section, we show that they also anticipated the work of Schotch and Jennings, though in a way that is not at all obvious.

Jónsson and Tarski observed in their first paper on Boolean algebras with operators Jónsson and Tarski (1951) that additive operators are not closed under composition. Let us see why this is so. The n-place projection operators π_i^n are defined on a set L by the equation $\pi_i^n(\vec{a}) = a_i$, where $\vec{a} \in L^n$; these operators on a distributive lattice are monotone and additive. If $F(\vec{x})$ is an n-place additive operator on a distributive lattice, then the unary operator G defined by

$$G(\vec{x}) = F(\pi_1^n(\vec{x}), \pi_1^n(\vec{x}), \ldots, \pi_1^n(\vec{x}))$$

is the diagonalization of F. As we have seen above in §9.3, G may not be additive.

Let X be a fixed set, and Φ a class of functions defined on X. Then the *closure of Φ under composition* is the smallest class Ψ of functions satisfying

1. $\Phi \subseteq \Psi$;

2. If, for some m and n, $f \in \Psi$ is a function on X^m to X and $g_1, \ldots, g_n$ are functions on X^n to X, then $f[g_1, \ldots, g_n] \in \Psi$.

Theorem 9.6.1 is an extension of a theorem of Jónsson and Tarski (1951) characterizing the closure of the additive functions under composition.

Theorem 9.6.1. *Let L be a bounded distributive lattice, and $\Diamond$ a normal function from L^n to L. The following conditions are equivalent:*

1. *The function $\Diamond$ belongs to the closure under composition of the class of monotone additive functions on L;*

2. There is an n-place monotone additive function ∇ on L and projection functions $\pi_1, \ldots \pi_n$, so that

$$\Diamond(a_1, \ldots, a_n) = \nabla(\pi_1(a_1, \ldots, a_n), \ldots, \pi_n(a_1, \ldots, a_n)),$$

where each π_k is π_i^n, for some i, $1 \le i \le n$;

3. There is a vector $\vec{k}$ of non-negative integers, where $|\vec{k}| = n$, and a monotone additive function ∇ on L so that for all $a_1, \ldots, a_n \in L$,

$$\Diamond(a_1, \ldots, a_n) = \nabla(a_1^{k_1}, \ldots, a_n^{k_n});$$

4. The function $\Diamond$ is a weakly multimodal operator on L.

Proof. The equivalence of the first two conditions follows by Theorem 1.10 of Jónsson and Tarski (1951).

The equivalence of the second and third conditions follows from the fact that the family of monotone additive functions is closed under permutation of arguments. That is to say, if $F(\vec{x})$ is a monotone additive function, then so is the function G defined by $G(\vec{x}) = F(\vec{x}^\sigma)$, where σ is any permutation of the elements of the vector of variables $\vec{x}$. Composing a monotone additive function with projection functions results in a function in which certain variables are identified, for example, a five-place function F might give rise to the three-place function $F(x_1, x_2, x_1, x_3, x_1)$. A permutation of the variables results in a new multimodal function $F'(x_1, x_1, x_1, x_2, x_3)$, where identical arguments are adjacent. This shows the equivalence with the third condition.

If the operator $\Diamond$ satisfies the third condition, then it satisfies the fourth, by Lemma 9.3.2. To show the converse, assume that $\Diamond$ is a weakly multimodal operator of type $\vec{k}$ on a lattice L, where $m + 1 = \sum \vec{k}$. By Theorem 9.4.4, the algebra $\langle L, \Diamond \rangle$ is isomorphic to the dual algebra $L' = L(\mathcal{R}(L))$ of its dual space. Let $R_\Diamond$ be the m-place relation on $\mathcal{R}(L)$. Define an m-place multimodal operator ∇ on L' by the definition:

$$x \in \nabla(A_1, \ldots, A_m) \Leftrightarrow \exists \vec{y_1}, \ldots, \vec{y_m} [R x \vec{y_1}, \ldots, \vec{y_m} \wedge \forall i \, [1 \le i \le n \Rightarrow y_i \in A_i]].$$

It follows immediately that the operator $\Diamond$ satisfies the second condition in the statement of the theorem. □

Theorem 9.6.1 shows that the family of weakly additive functions, appears in the work of Jónsson and Tarski, although only in an implicit form. What is missing from their paper is an axiomatic characterization of the class, a characterization that had to wait for the work of Jennings and Schotch.

9.7 The Finite Model Property and Decidability

A family of algebras is defined to be a *variety* if it is defined by a set of equations. For example, distributive lattices, Boolean algebras, multimodal algebras and weakly multimodal algebras of type $\vec{k}$ all form varieties, since they are characterized by axiom sets consisting of equations.

A variety V is *generated by its finite members* if the smallest variety containing all the finite algebras in V coincides with V itself. Equivalently, a variety is generated by its finite members if and only if the following holds: if an equation $\sigma = \tau$ is not universally valid in V, then there is a finite algebra $\mathfrak{A}$ in V so that $\sigma = \tau$ can be refuted in $\mathfrak{A}$. The varieties of distributive lattices and Boolean algebras are generated by their finite members, in fact both varieties are generated by two-element algebras. The importance of this property lies in the fact that it often leads to decidability proofs.

Theorem 9.7.1. *Let V be a variety of algebras with the following two properties:*

1. *The identities of V can be computably enumerated;*

2. *V is generated by a computably enumerable set of finite algebras.*

Then it is decidable whether or not an equation is valid in V.

Proof. Given an equation $\sigma = \tau$, we can enumerate the identities of V, and at the same time the set of finite algebras that generates V. Eventually, we either find $\sigma = \tau$ in the list of theorems of V, or a finite algebra in which it can be refuted. □

The preceding theorem is of course just an algebraic version of the familiar result in the theory of nonclassical logics proving decidability from the finite model property.

Theorem 9.7.2. *The varieties of multimodal and weakly multimodal algebras of type $\vec{k}$ are generated by their finite members.*

Proof. The proof that the variety of multimodal algebras is generated by its finite members follows by an algebraic version of the filtration method from the theory of modal logic (Blackburn and van Benthem, 2007, p. 29f). Since the proof is a routine adaptation of standard results, we omit it here.

Let $\sigma = \tau$ be an equation that is not valid in the theory $V^{\vec{k}}$ of weakly multimodal algebras of type $\vec{k}$, where $|\vec{k}| = n$, and $\Diamond$ the weakly multimodal operator of type $\vec{k}$ in the theory. By Theorem 9.6.1, any algebra in $V^{\vec{k}}$ is derived from an algebra in the variety V_m of all m-ary multimodal algebras, where $m = \sum \vec{k}$, by the $\vec{k}$-collapse operation. Hence, the variety $V^{\vec{k}}$ can be exactly embedded in the variety obtained from V_m by adding the definition:

$$\Diamond(a_1, \ldots, a_n) =_{\text{df}} \nabla(a_1^{k_1}, \ldots, a_n^{k_n}).$$

Now let $\sigma' = \tau'$ be the equation derived from $\sigma = \tau$ by replacing the $\Diamond$ operation by the corresponding definiendum in V_m. Since V_m is generated by its finite members, there is a finite multimodal algebra in V_m so that $\sigma' = \tau'$ fails in it. Applying the definition above to this algebra results in an algebra in $V^{\vec{k}}$ in which the original equation $\sigma = \tau$ fails. Hence, $V^{\vec{k}}$ is also generated by its finite members. □

Corollary 9.7.3. 1. *The equational theories of the varieties V_n and $V^{\vec{k}}$ are decidable;*

2. The logic K_n of Schotch and Jennings is decidable.

Proof. The first part of the Corollary follows by Theorems 9.7.1 and 9.7.2. The second part follows because the logic K_n is the logical counterpart of the variety $V^{\langle n \rangle}$. That is to say, the proof that $V^{\langle n \rangle}$ is generated by its finite members can be translated directly into a proof that the logic K_n has the finite model property. □

9.8 Bibliography

Patrick Blackburn and Johan van Benthem. Modal logic: a semantic perspective. In Patrick Blackburn, Johan van Benthem, and Frank Wolter, editors, *Handbook of Modal Logic*, pages 1–84. Elsevier, 2007.

Bryson Brown. Preservationism: A short history. In Dov M. Gabbay and John Woods, editors, *Handbook of the History of Logic, Volume 8: The many valued and nonmonotonic turn in logic*, pages 95–127. North-Holland Elsevier, 2007.

B.A. Davey and D. Duffus. Exponentiation and duality. In I. Rival, editor, *Ordered sets*, pages 43–95. Reidel, 1982.

B.A. Davey and H.A. Priestley. *Introduction to lattices and order*. Cambridge University Press, Cambridge, 1990.

Isidore Fleischer. "Kripke semantics" = algebra + poetry. *Logique et Analyse*, 27: 283–295, 1984.

B. Jónsson and A. Tarski. Boolean algebras with operators. Part I. *Amer. J. Math.*, 73: 891–939, 1951.

B. Jónsson and A. Tarski. Boolean algebras with operators. Part II. *Amer. J. Math.*, 74: 127–162, 1952.

Saul A. Kripke. Semantical analysis of modal logic I: Normal modal propositional calculi. *Zeitschrift für mathematische Logik und Grundlagen der Mathematik*, 9: 67–96, 1963.

Kam Sing Leung and R.E. Jennings. Polyadic modal logics and their monadic fragments. In Peter Schotch, Bryson Brown, and Raymond Jennings, editors, *On Preserving: Essays on Preservationism and Paraconsistent Logic*, pages 61–83. University of Toronto Press, 2009.

S. Mac Lane. *Categories for the Working Mathematician*. Springer-Verlag, 1971. Graduate Texts in Mathematics 5.

H.A. Priestley. Representation of distributive lattices by means of ordered Stone spaces. *Bull. London Math. Soc.*, 2:186–190, 1970.

H.A. Priestley. Ordered sets and duality for distributive lattices. *Annals of Discrete Mathematics*, 23:39–60, 1984.

Alasdair Urquhart. Weakly additive algebras and a completeness problem. In Peter Schotch, Bryson Brown, and Raymond Jennings, editors, *On Preserving: Essays on Preservationism and Paraconsistent Logic*, pages 33–47. University of Toronto Press, 2009.

Chapter 10

From Consequence Relations to Reasoning Strategies

M. Bryson Brown[1]

10.1 Introduction

Henry Kyburg often pointed out, in (Kyburg, 1970) and elsewhere, that while ∧-I is *truth-preserving*, it is also *probability-degrading*. Worse, while we can have extremely strong evidence for each individual sentence in a finite but inconsistent set, a sequence of ∧-I steps turns the set into a single, contradictory sentence. This may not trouble some: dialethiests have argued that we should accept certain contradictions. But the applications they aim at differ from ours in important ways, and their project tolerates contradictions not so much for themselves as for the sake of the formal system they belong to. Our concerns are more practical; since we see no practical use for such contradictions in the systems they belong to, we prefer to avoid them rather than deal with them by adopting a logic that renders them safe. On our view, individually consistent members of some inconsistent sets of sentences often play useful roles in reasoning, while the contradictions that follow when we aggregate those sentences freely are useless.

Limiting aggregation is a simple strategy for avoiding the logical trivialization of inconsistent sets of sentences. We also hold that it leads to illuminating models of how scientists and others have managed to reason effectively with inconsistent premises.

10.2 Forcing and Variations

Logics that weaken a base consequence relation in this way are called *weakly aggregative* (Schotch and Jennings, 1989). The Schotch-Jennings *level* of a premise set Γ, $l(\Gamma)$, is defined as the least n for which there is a consistent n-partition of Γ, else ∞. Using $[\vdash$ for the Schotch-Jennings forcing relation and assuming $l(\Gamma) = n$,

---

[1]Department of Philosophy, University of Lethbridge, brown@uleth.ca

$\Gamma[\vdash \alpha$ iff for every n-covering of Γ G, $\exists \gamma \in G : G \vdash \alpha$.

For level n, the *aggregation* forcing imposes on Γ is determined by the rule of $2/n + 1$ introduction: where

$2/n + 1(\alpha_0, ...\alpha_n) = \bigvee_{0 \leq i \neq j \leq n}(\alpha_i \wedge \alpha_j)$,
$\Gamma[\vdash \alpha_0, ...\Gamma[\vdash \alpha_n \Rightarrow \Gamma[\vdash 2/n + 1(\alpha_0, ...\alpha_n)$ (Apostoli and Brown, 1995).

This is the strongest purely *formal* principle of aggregation applying to premises with level n that avoids trivialization.

As a logical *organon* in Aristotle's sense, forcing is clearly an improvement: we retain the base logic so long as the premises are consistent, while avoiding the risk of trivialization if we should stray into inconsistency (a hard thing to avoid when dealing with large collections of commitments, even if our methods are very reliable). Where classical and other 'explosive' logics simply tell us to go out and get an acceptable premise set before trying to reason with it, forcing allows us to work more constructively with the premises we actually have.

However, quantifying across every n-covering of Γ weakens aggregation drastically. Even at level 2, the lowest level that lacks classical $\wedge$-I and limits aggregation to $2/3$: this aggregative rule produces the disjunction of pairwise conjunctions amongst any three sentences in Γ. $2/3$ (and $2/n + 1$ in general) is *aggregative*: in general, the results of applying this rule don't follow from the individual sentences the rule was applied to. But the resulting aggregation is quite weak, and that weakness is essential to level preservation: conjoining consistent subsets G of an inconsistent set of sentences Γ doesn't produce contradictions, but (in general) it increases the *level* of Γ.

Reflecting on Kyburg's lottery paradox makes this point very clear: consider a lottery case in which Γ includes the sentences 'ticket i' will not win, for $i = 1, ..., 1000$, together with the sentence 'for some i, ticket i will win'. Every conjunction of a 1000-membered subset of Γ is consistent. But the level of the set of such conjunctions is 1001, while the level of the original set is just 2. Kyburg's probabilistic concerns here also count firmly against aggregating our commitments to the limits of consistency: in a lottery case, the probability of the resulting sentences is $1 - p$(ticket n will not win). Anyone concerned (as Kyburg was) to avoid commitments to claims with very low probabilities will be reluctant to go this far simply because the results are still individually consistent.

We can combine our sense that conjoining some sentences in an inconsistent collection of sentence is acceptable with the concern that conjunction introduction is generally probability degrading. In such an approach to inconsistency management, we generalize 'level' along probabilistic lines: a *probabilistic level* can be defined in terms of the minimum acceptable probability of a accepted sentence, allowing aggregation that ensures probabilities at or above that threshold probability.(Brown, 1999) But such limits don't depend strictly on the *logical relations* of the sentences in question.

Another response to this weakness is to restrict the range of coverings applied to determine the consequences of Γ. Some logically-grounded measures can be taken along these lines. If we define a 'harmless bystander' as a sentence that appears in every maximal consistent subset of Γ, we can require all such 'harmless bystanders' be included in every cell of every admissible covering.(Payette, 2015), since they make no contribution to the inconsistency of Γ.

10.2. FORCING AND VARIATIONS

We could also require that every cell of an admissible covering include a maximal consistent subset of Γ. Where $l(\Gamma) = n$, there is at least one consistent n-covering of Gamma; the members of any such covering can be extended (in general, in many ways) to maximal consistent subsets of Γ. If we insist that only such coverings be considered, then the consequences of Γ will be the sentences that follow from every such *maximal* n-covering of Γ.

Finally, we can think a little more broadly about the contents of our cells: rather than merely add certain sentences from Γ itself, we can consider the *consequences* of the maximal consistent subsets of Γ. Since any consistent subset of Γ can be extended to some maximal consistent subset of Γ, the intersection of the consequences of the maximal consistent subsets of Γ can be consistently added to any cell of a consistent n-partition of Γ. The result is what Payette calls 'M-forcing', which combines Rescher and Manor's weaker 'shared consequences of maximal consistent subsets' rule for determining the content of an inconsistent set of sentences.(Payette, 2015; Rescher and Manor, 1970, p. 5),

All these are *formal* in the sense that the underlying consequence relation $\vdash$ determines both the maximal consistent subsets of Γ and the 'harmless bystanders' of Γ. But all of these consequence relations drastically weaken the inferential links between the sentences of Γ. Even if we go beyond what the logical relations between the sentences in Γ tell us by selecting a unique partition of Γ into a collection of maximal consistent subsets of Γ, and then close each cell of the partition under $\vdash$, the result is just the *union* of n theories in the underlying logic. From a logical point of view, we can call this union an 'n-theory' if we like, but a skeptical listener might justifiably dig in her heels and insist that it is *better* described as a union of n *proper* theories.

The limits we encounter here arise from the standard notion of a *consequence relation* as a relation between ensembles of sentences and their (unique) *closure* under the consequence relation. On the standard reading, the closure of an ensemble under a consequence relation is then taken to constitute the *commitments* that 'follow from' or are *undertaken* when someone makes a commitment to the ensemble's members. Consequence relations of this kind are taken to provide normative models of reasoning, in which reasonings that cite the members of an ensemble E as premises and proceed by applying inference rules that guarantee we will arrive at a member of the closure of E as a conclusion are endorsed as correct. We might call this the *standard logical model* of reasoning.

But this model is clearly an *idealization*: human reasoning doesn't resemble this model closely at all. As an idealization, it is useful and justifiably influential, but real reasoning emerges from a much more complex engagement with the available premises. On one hand, many inferences endorsed by the standard model are never actually made, as Harman has emphasized: many logically correct inferences (such as long strings of ∧ or ∨-I steps) are never made in 'real life,': they are useless to us, despite meeting the criteria of correctness. More substantively, the inferences typical of rich and complex contexts are goal-driven, often exploratory, and generally focused on arriving at conclusions of particular interest, by drawing on particular combinations of premises seen as relevant to the questions reasoners seek answers to. Rarely if ever do we find the sort of careful, systematic and (at least sometimes) consistent treatments that familiar logical models of consequence capture, and (more importantly) such treatments don't emerge until *after* many particular inferences have been explored and found to be useful.(Brown and Priest, 2004, 2015)

Of course reasoning in the wild is very different from reasoning that has been collected, taxonomized and taxidermied to produce an exhibit for philosophical reflection. Outside of logical and formal mathematical contexts, the generalized application of rules of inference to all sentences appealed to in the course of reasoning is imposed only in the course of post-hoc reconstruction of both the premises and the inferences that drew on them; but these reconstructions should be understood not as mere *accounts* of the original reasoning, but instead as *reinterpretions*.

In this paper I present a systematic approach to reasoning that can be useful in contexts where premises needed at some points in the course of reasoning are inconsistent with premises applied at later points; their application along the way can nevertheless be rightly regarded as *reliable* in the way that Bohr's orbiting, non-radiating electrons proved to be a reliable basis for predicting many features of the hydrogen spectrum.(Brown and Priest, 2015)

10.3 Chunk and Permeate

The 'Chunk and Permeate' ($C\&P$)(Brown and Priest, 2004) inference strategy uses a *permeation relation* to strengthen the links between the members of inconsistent premise sets. The upshot is a systematic approach to building models of reasoning with inconsistent premises that avoids the derivation of contradictions, but relies on real inferential links between incompatible commitments: broadly described, this strategy allows specific *kinds* of conclusions inferred from certain premises to be applied in further reasonings which invoke contrary premises, leading ultimately to useful conclusions.

A $C\&P$ structure on Γ, $\wp$, is a 3-tuple $\langle P, \rho, i_0 \rangle$ where:

i. P is a consistent covering of Γ, with elements $\gamma_1, ... \gamma_n$

ii. ρ is a *permeation relation*, a function from $I \times I$ into L. All and only the sentences in $\rho(i,j)$ are allowed to permeate from γ_i to γ_j.

iii. i_0 is the label of the *designated chunk*, where we draw our conclusions.

The $C\&P$ consequences of Γ relative to a $C\&P$ structure $\langle P, \rho, i_0 \rangle$ are produced by a series of closure and permeation steps. Where γ_i is the ith cell of P, γ_i^n is defined recursively:

- $\gamma_i^0 = CL\gamma_i$
- $\gamma_i^{n+1} = CL(\gamma_i^n \cup \bigcup_{j \in I}(\gamma_j^n \cap \rho(j,i)))$

We define the $C\&P$ consequences of such a structure as the closure of a particular cell (the 'output' chunk) under this recursion:

10.4. LEVEL-PRESERVATION IN C AND P

- $\Gamma \Vdash_\wp \alpha$ iff $\exists n, \alpha \in \gamma_{i_0}^n$.

$C\&P$'s successive closure and permeation operations ensure that the premises available in each chunk can contribute to the conclusions drawn in the output chunk. But $C\&P$ structures take us a long way from the standard concept of a consequence relation: in particular, $\bigcup \gamma_n^i$ doesn't determine the contents of $\gamma_{i_0}^n$ all by itself. So this 'consequence relation' is *not* a relation between ensembles of sentences and the sentences that follow from them; if we accept that logical consequence must be such a relation, $C\&P$ structures do not produce logical consequence relations.

A pragmatic element is involved here: *logical* criteria alone don't tell us which covering of Γ or which permeation relation is the 'right one'; answers to these questions turn on more than just the sentences in $\bigcup \gamma_n^i$. In applications of the $C\&P$ model, the coverings and permeation relations proposed are constrained pragmatically, by the actual inferential practices we are applying $C\&P$ to model. A *well-designed $C\&P$* structure will allow the inferences actually made in the course of reasoning with the sentences in $\bigcup \gamma_n^i$, such as the old calculus or Bohr's model of the hydrogen atom, while ruling out inferences that were not accepted (and especially those that would have led to trivialization).

So $C\&P$ is not a formal consequence relation linking ensembles premises to their closure under the relation. Instead, it offers a general *strategy* for building inconsistency-tolerating *inferential machinery*. As such, it aims to strike a balance between the need to keep logical models close to the phenomena (i.e. to how people actually reasoned, including the premises they relied on and the inferences they made from them) and the philosophical aim of providing a workable, *explicit* reconstruction of how their conclusions could be reached while avoiding both arbitrariness and triviality.

10.4 Level-preservation in C and P

$C\&P$ structures are not intrinsically level-preserving. The conclusions that follow from a $C\&P$ structure S are determined by the contents of the output chunk at the limit of our cycle of closing all chunks under $\vdash$ and then allowing sentences that belong to $\rho(i,j)$ to permeate from σ_i to σ_j for all chunks i,j. Nothing in this definition guarantees that the level of the union of chunks $\bigcup_1^n \sigma_i$ will be preserved. Of course failure to preserve Post-consistency of the output chunk is as disastrous for a $C\&P$ structure as it is for a standard logical closure operation. But, while avoiding this sort of disaster is a sine qua non for building a useful $C\&P$ structure, nothing in the nature of these structures protects us from it. From this perspective, $C\&P$ is not paraconsistent at all: it's no harder to trivialize the output chunk of a $C\&P$ structure than it is to trivialize premise sets with the 'help' of a non-paraconsistent consequence relation.

Of course a *successful* (or even minimally useful) $C\&P$ structure preserves the Post-consistency of its output chunk, while allowing an inconsistent collection of premises, applied in the separate chunks, to contribute to the contents of the output chunk. And so long as all the individual chunks remain consistent, the level of $\bigcup_i = 0^i = n\sigma_i$ must be less than or equal to n.

However, under the right conditions, it's possible to avoid trivializing the output chunk even if level-preservation for the set of all the sentences in the structure, $\bigcup_i = 0^i = n\sigma_i$, fails. The milder kind of case where this can occur involves a $C\&P$ structure with more than $l(\bigcup_i = 0^i = n\sigma_i)$ chunks; the cycle of permeation and closure can then lead to an increase in level while preserving the consistency of each chunk. A toy example can be built using three chunks, $\sigma_0, \sigma_1, \sigma_2$ and the classical $\vdash$: Let $\sigma_0, \sigma_1, \sigma_2$ begin with the contents 'p, q', '$\neg p$' and '$(q \to p)$', respectively, while $\rho_{1,2}$ includes all wffs that don't include 'p', $\rho_{2,3}$ includes the negative literals of L and $\rho_{2,3}$ includes atoms of L only. At the start, $l(\bigcup \sigma_0, \sigma_1, \sigma_2) = 2$. But at the first 'close and permeate' step, 'q' (along with all its consequences except those including 'p') is added to σ_1, and '$\neg p$' is added to σ_2. At this point, $l(\bigcup_i = 0^i = n\sigma_i) = 3$. No further increase in $l(\bigcup_i = 0^i = n\sigma_i)$ occurs at subsequent stages of the recursion.

A more extreme kind of case can also arise, in which one or more chunks trivialize, but restrictions on ρ prevent that trivialization from spreading to the output chunk. (Obviously enough, preventing trivialization of the output chunk in such cases requires tight restrictions on permeation from the trivial cells.) A toy model of this sort of case has three chunks, $\sigma_0, \sigma_1, \sigma_2$; at the outset, σ_0 includes 'p', σ_1 contains '$\neg p$' and σ_2 is empty. The permeation relation $\rho_{0,1}$ is empty, while $\rho_{0,2}$ and $\rho_{2,0}$ include all the atoms of L and $\rho_{1,2}$ includes the literals of L. If we make σ_0 the output chunk, the consequences of this structure are the sentences in the closure of the atoms of L under $\vdash$.

However, when it comes to actual applications of $C\&P$, such cases are better *modeled* by $C\&P$ structures that do preserve level. For every chunk σ_i that trivializes at some point during the recursion, we delete σ_i from the structure while adding all the sentences in $\rho(\sigma_i, \sigma_j)$ to each σ_j as part of its initial content.

The upshot is that level still plays a significant role for $C\&P$, though this role is somewhat different from the role it plays in forcing. Rather than restrict our inference rules to ensure preservation of level, we aim to produce a *structure* that prevents inconsistencies from arising in any chunk. The upshot is that a well-designed structure preserves the level of the set of sentences that populate its chunks. In a sense, this is a generalization of our everyday epistemic efforts to build a consistent account of the world, given that we cannot be certain we have started with a consistent body of commitments: we don't simply assume that our starting point is consistent; instead, we avoid *trivialization* by restricting how our premises are aggregated, how they are reasoned with and how the results of our reasonings are combined in further reasonings. The resulting models of reasoning can tolerate inconsistency without logical heterodoxy, positioning $C\&P$ somewhere in between the two responses to inconsistency in science Norton discusses in (Norton, 2002): there, Norton distinguishes logical from content-driven approaches to dealing with inconsistency in science, arguing against the logic-driven approach that the content-driven approach allows us to cope with inconsistency without changing "something as fundamental as *logic*."(Norton, 2002, p. 193)

10.5 Applications

The $C\&P$ strategy has been applied to model reasoning in the old calculus (Brown and Priest, 2004), Bohr's model of the hydrogen atom (Brown and Priest, 2015) and the Dirac delta function(Benham et al., 2014). In these cases, our aim has been to do more than provide a reasonable model of how these mathematicians and physicists could have avoided trivialization while working with their inconsistent commitments; we have also aimed at producing models that supported particular arguments that draw directly on on those commitments. For example, in (Brown and Priest, 2004) we proposed a $C\&P$ structure LN= $\langle \{\Sigma_S, \Sigma_T\}, rho, T\rangle$. Σ_S includes a second-order theory of the reals (or a sufficiently strong sub-theory) along with an equation which defines a derivative $Df = \lambda x(f(x+\delta x) - f(x)/\delta x$ and an equation specifying $\delta x \neq 0$. The second chunk includes neither of these equations, but does include $(x)\delta x = 0$. The permeability function $\rho(T, S)$ is empty, while $\rho(S, T)$ is limited to equations $Df = g$ where neither f nor g contains D, f is a λ-term with no occurrences of δ, and g is of the form $\lambda x(h+p)$ where h does not contain δ and p is a polynomial of powers > 0 of δx.

The result is that the key move in the old calculus, in which we rely the assumption that $\delta x \neq 0$ to calculate a value for Df that doesn't have δx in the denominator and then suddenly declare that $\delta x = 0$, does not commit us to the obvious contradiction $\delta x \neq 0 \wedge \delta x = 0$: the quantity is only allowed to depart after the assumptions that required $\delta x \neq 0$ have been explicitly set aside.

One further aim of work with $C\&P$ has been to avoid imposing a specific logic as part of these models of constrained reasoning from inconsistent premises. Nuel Belnap once asked me how we could tell what logic was being used by a working scientist– and the question is a hard one to answer. Actual reasoning can often be interpreted using a wide range of logics: the inference rules accepted in different logics generally overlap, and enthymemes and logical short-cuts are common in scientific papers. Making a related point, John Norton distinguishes what he calls 'content-driven' and 'logic-driven' control of logical anarchy (Norton, 2002, p. 191f), arguing in favour of content-driven approaches. Norton suggests that the content-driven approach can help us to separate the "pathological" conclusions that follow from an inconsistent theory T from those that would be *retained* in a (satisfactory) consistent replacement for T. For example, Norton suggests that symmetry considerations support the assumption that inconsistent results for forces acting on a test-body in an infinite Newtonian universe with non-zero average density should be dropped in favour of the assumption that the net force acting on a test body due to gravitational influences on the cosmological scale is zero. Similarly, $C\&P$ offers an approach to coping with inconsistency that does not turn on altering the underlying *logic*. $C\&P$ is certainly 'content-driven': it succeeds only if the division of the contents combined with the permeation relation yields conclusions that are 'wanted on the journey' (and can serve as guides to the evaluation of a satisfactory, consistent replacement theory). But $C\&P$ occupies a middle ground here, emphasizing closer attention to actual reasoning used by scientists, but applying a general logical plan combining separation into chunks with permeation to construct a model of how trivialization was avoided.

Closure under a formal consequence relation provides a systematic account of the cognitive commitments that follow from accepting some collection of premises. But this point of view is one we can only hope to achieve at the end of the day: the epistemological process that gives rise to candidates for this kind of acceptance is always tentative and exploratory. Locally successful reasoning patterns are identified and applied long before the emergence of systematic theories in which they are later embedded (and re-embedded).

$C\&P$ structures model patterns of reasoning without commitment to any kind of logical closure of all the sentences we appeal to in the course of reasoning. When the conclusions we reach by these methods turn out to be reliable, we can work to extend and systematize the patterns of reasoning involved, perhaps eventually discovering how to capture the key results in a consistent way. On this reading, $C\&P$ structures provide systematic models of this process *in medias res*.

At the end of the day, when we're lucky, we wind up with a systematized collection of premises that we can rely on in a wide range of circumstances and even consider accepting for closure under a consequence relation. But the aim of $C\&P$ (at least as I see it) is to provide an approach to representing the exploratory, pre-systematic reasonings that sometimes lead us to such systematically successful theories.

10.6 Bibliography

Apostoli, P. and Brown, B. "A Solution to the Completeness Problem for Weakly Aggregative Modal Logic," *The Journal of Symbolic Logic* **60**, 3: 832-842, 1995

Benham, Mortensen, C. and Priest, G. "Chunk and Permeate III: The Dirac Delta Function" *Synthese* **191**, 13: 3057-3062, 2014.

Brown, B. "Adjunction and Aggregation" *Nous* **33**, 2: 273-283, 1999.

Brown, B. and Priest, G. "Chunk and Permeate" *The Journal of Philosophical Logic* **33**, 4: 379-388, 2004.

Priest, G. and Brown, B. "Chunk and Permeate II: the Bohr Atom" *European Journal for the Philosophy of Science* DOI 10.1007/s13194-014-0104-7, January, 2015.

Kyburg Jr., H.E. "Conjunctivitis" 55-82 in (Swain, 1979).

Meheus, J. (ed.) *Inconsistency in Science* Kluwer Academic Publishers, The Netherlands, 2002.

Norton, J. D. "A Paradox in Newtonian Cosmology" 412-20 in (Meheus, 2002). 1993.

Payette, G. "Getting the most out of inconsistency" *Journal of Philosophical Logic* **44** (5) 573-592 2016.

Rescher, N. and Manor, R. "On inference from inconsistent premises" *Theory and Decision* 1, 179-217 1970.

Schotch, P. K., Brown, B. and Jennings, R. E. (eds.) *On Preserving: Essays on Preservationism and Paraconsistent Logic* University of Toronto Press, Toronto, 2009.

Schotch, P. K. and Jennings, R. E. "On Detonating" 306-327 in (Sylvan et al., 1989).

Swain, M. *Induction, Acceptance and Rational Belief* D. Reidel, Dordrecht, 1970.

Sylvan, R., Priest, G. and Norman, J., (eds.) *Paraconsistent Logic: Essays on the Inconsistent* Philosophia Verlag, München, 1989.

Chapter 11

Type Raising:

Schotch on the Prisoner's Dilemma and Deontic Logic

Gillman Payette[1]

11.1 Introduction

In this paper I look at two issues which are related by Schotch's work in deontic logic. The first is his diagnosis of what is paradoxical in the prisoner's dilemma. The second is deontic semantics. What connects these two areas is Schotch's approach to social choice, and within that the topic known as 'type raising'. The topic of type raising deals with how to 'raise' the 'type' of binary relations from individuals to sets of individuals. What this paper lacks in comments directly about Schotch's work, it makes up for in inspiration gained from the man himself. My overall goal in this paper is to show how important the scheme used in type raising is to the analysis of value theory.

In section 11.2 I explain type raising and fix some notation for latter discussion. In section 11.3 I deal with Schotch's account of the prisoner's dilemma. Then I turn to deontic logic in section 11.4.

Schotch argues that the reason why the prisoner's dilemma is paradoxical is because it suggests that we should act in a way that we think is crazy. I argue that although Schotch's justification relies on aspects of his formulation of the dilemma that are unnecessary in general, his conclusion is sound. However, in order to maintain his conclusion I argue further that either he does hold the Pareto principle in high regard, or ought to hold it in high regard.

When I turn to deontic semantics, I offer a fairly general account of obligation and permission which deals with actions and bringing about propositions: these are both forms of 'ought to do'. The account is general enough to encompass stit semantics of various kinds and the conceptions of deontic logic from game theory. All of these accounts are based on an axiological account of what ought to be done; i.e., doing the best thing according to some notion of value. The assumption is that value is accorded

---

[1]Department of Philosophy, University of British Columbia, gpayette@dal.ca

to social states via a preference relation and that relation is raised to compare sets of states. I then consider how the way one raises the preference relation affects the semantics of the deontic terms. I first consider how it affects the relationship between axiological ideals and what ought to be done, and then how it could affect the limited reasoning capacities of groups.

11.2 Type Raising

I will use $\leq$ and $<$ as arbitrary relations (or the usual orderings on numbers) where the latter is the strict relation correlated to the first, i.e., $x \leq y$ but $y \nleq x$. In these relations, $x \leq y$ means that x is less than or equal to y. The idea of type raising, generally, is to take a relation on objects and turn it into a relation on sets of those objects. This process is important in rational decision theory when it comes to making decisions under ignorance or complete uncertainty, cf. Resnik (1987).[2] In that field, the relation in question is a *preference relation*. In what follows xP_iy will mean 'i strictly prefers x to y', xI_iy for 'i is indifferent between x and y', and xR_iy for 'x is at least as good as y for i' which can be thought of as the disjunction of P_i and I_i. Alternatively, xP_iy can be thought of as the conjunction of xR_iy and not-yI_ix, or, equivalently, xR_iy and not-yR_ix. The standard assumptions about the weak preference relation R_i are the following:

Definition 1. A preference relation R_i is a relation on some set of entities W that is a total pre-order, i.e., it is

- Transitive $\forall x, y, z \in W((xR_iy \,\&\, yR_iz) \supset xR_iz)$,
- Reflexive $\forall x \in W(xR_ix)$, and
- Total $\forall x, y \in W(xR_iy \vee yR_ix)$.

The set W can be of anything. It could be a set of commodities, it could be a set of people, or it could be a set of possible worlds. Since P_i is strict, it is irreflexive, but it is transitive. I_i is an equivalence relation since it is also symmetric. As is usually done, the properties

$$\forall x, y, z \in W((xP_iy \,\&\, yI_iz) \supset xP_iz),$$

and

$$\forall x, y, z \in W((xI_iy \,\&\, zP_iy) \supset zP_ix)$$

will also be referred by the name 'transitivity'.

For any set $X \subseteq W$ the top and bottom members of the set relative to R_i are defined as follows.

Definition 2. Let $X \subseteq W$, and $\leq$ be a relation on W, then

---

[2]Along these lines another former student of Peter's has written a Master's thesis on type raising and decisions under complete uncertainty which compares the philosophical literature to the economic literature on the subject, Phang (2012).

11.2. TYPE RAISING

$$\max(X) = \{\, x \in X : \forall x' \in X \; x' \leq x \,\}, \text{ and}$$

$$\min(X) = \{\, x \in X : \forall x' \in X \; x \leq x' \,\}.$$

It is important to notice that these functions may return empty sets when the set X is infinite. Often in the economics literature the sets considered are finite, but since I will later be dealing with semantics for deontic logic, I will not make that assumption.

From a preference relation on *members* of W one can derive a preference relation on *subsets* of W. However, there isn't a unique way to do this, but there are reasons for and against the various ways of doing this. I will look at some of the derivations that appear in the literature; here is a list:

- For $X, Y \in \mathcal{P}(W)$, $X \leqslant_i^* Y$ (Y is at least as good as X for i) iff

(v) $\exists y \in Y \forall x \in X$ s.t. $yR_i x$ (van Fraassen, 1973)

(J) $\forall x \in X \exists y \in Y$ s.t. $yR_i x$ (Jennings, 1974; Lewis, 1973)

(H) $\forall y \in Y \forall x \in X, yR_i x$ (Horty, 2001)

(S) $\forall x \in X \exists y \in Y$ s.t. $yR_i x$ and $\forall y \in Y \exists x \in X$ s.t. $yR_i x$ (Schotch, 2016)

(R) $\forall x \in \min(X) \forall y \in \min(Y), yR_i x$ (MaxiMin)

From the list above, replacing the $*$ in the relation on sets with one of the letters from the list: v, J, H, or S, gives a type raised relation $\leqslant^*$ between sets. I usually omit the reference to the agent i. From the non-strict version $\leqslant^*$, a strict version $<^*$ can be defined by $X <^* Y$ iff $X \leqslant^* Y$ and $Y \not\leqslant^* X$. But unlike in the usual case, the definitions of the strict version of a type raised relation can be rather complicated.

Let's notice some properties of and connections between, these conditions. Most of these have been illustrated in Schotch (2016). First, each of the schemes mirrors the relation R_i when comparing unit sets, i.e.,

$$\{x\} \leqslant^* \{y\} \iff yR_i x$$

for each $*$. When sets X, Y are not comparable with respect to $\leqslant^*$, I will write $X \perp^* Y$. Some immediate connections to notice are that H implies v, S and J, that the condition v implies J, and condition S implies J. Indeed, condition S is a combination of J and what Schotch calls the 'Onto' condition since it makes the R_i relation restricted to the sets X and Y almost a surjective function from Y to X—it is just missing that each element in Y is mapped to a unique element of X.

There is also a connection between the strict parts of J and v. Note the following fact (see Schotch (2016) and Phang (2012)):

Observation 1. *$X <^v Y$ is equivalent to $\exists y \in Y \forall x \in X, yP_i x$.*

Proof. Assume $X <^v Y$, then by definition $X \leqslant^v Y$ and $Y \not\leqslant^v X$. Unpacking the definitions we have: (1) $\exists y \in Y \forall x \in X$ s.t. $yR_i x$ and (2) $\forall x \in X \exists y \in Y$ s.t. $yP_i x$ since R_i is total. There is a $y \in Y$ at least as good as every $x \in X$, call that element y.

Suppose $x \in X$. There are either $x' \in X$ such that $x'P_i x$ or not. Suppose not. Then either $yI_i x$ or $yP_i x$. If the second, then y is strictly better than every x by transitivity. If the former, then by (2) there is a y' preferred to x, and then $y'P_i y$ by transitivity. Then y' is preferred to every $x \in X$. Thus, $\exists y \in Y \forall x \in X, yP_i x$. The converse direction is immediate. □

Since $X \prec^v Y$ is equivalent to $\exists y \in Y \forall x \in X, yP_i x$, $Y \not\leq^J X$ iff $X \prec^v Y$. And another connection between S and v: $X \not\leq^S Y$ iff either $Y \prec^v X$ or $\exists y \in Y \forall x \in X$ s.t. $xP_i y$. That is, Y fails to be at least as good as X (according to scheme S) iff either X is preferred to Y (according to v) or there is something in Y that is worse than every element of X.

The strict version of $X \leq^H Y$ is to say that every element of Y is at least as good as every element of X but also that there is an element of X strictly worse than some element of Y according to R_i—the preference relation. Although, MaxiMin has received lots of attention since Rawls (1971), it doesn't matter for much in the discussions to come since for infinite sets, both functions can return empty sets. I only included it to stop people from asking 'what about MaxiMin?' A final thing to notice is that just because R_i is total—also referred to as 'complete'—doesn't mean that the type raised relation is as well. The J condition is complete, but the H, v and S conditions needn't give rise to complete relations over $\mathcal{P}(W)$. Since the H condition is the strongest condition, however, it will also be the *most* incomplete in the sense that every other condition will give at least as many comparisons between subsets of W as H does; most of the time they will give more comparisons since most subsets of an infinite set are infinite. Where H does have an advantage is in its robustness. For example,

Observation 2. *If $X \leq^H Y$, $X' \subseteq X$, and $Y' \subseteq Y$, then $X' \leq^H Y'$.*

The proof is immediate from the definition. But it is not immediate or otherwise for the other schemes. Now that I have discussed these preference relations, and how one can raise the relation from objects to sets of those objects, I can discuss what use can be made of these concepts. VERY IMPORTANT: there is a notational asymmetry between R_i and ≤. Whereas $xR_i y$ is read as 'x is at least as good as y' the notation $X \leq^* Y$ reverses that; it is read 'Y is at least as good as X—according to R_i'.

11.3 Schotch's Take on the Prisoner's Dilemma

The prisoner's dilemma (PD) has been a thorn in side of rational choice theory for a long time.[3] Here I will review it just to fix some notation. The set-up can be viewed in the following figure:

The suits indicate ways the world might turn out. Each player has two choices: cooperate or defect (C_i and D_i, respectively), although those glosses are not necessary for the scenario. Alice's choices can guarantee certain outcomes: if Alice chooses C_a, then the world will either be ♣ or ♠. Similarly, if Bob chooses D_b, then the world will

---

[3]Google can present one with the necessary references, so I won't.

11.3. SCHOTCH'S TAKE ON THE PRISONER'S DILEMMA

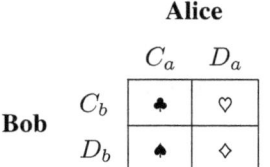

Table 11.1: The Prisoner's Dilemma

be either ♠ or ◊. Of course, if Alice chooses C_a and Bob choose D_b, then the world is guaranteed to turnout ♠. I will refer to those objects which represent they way the world might turn out as 'worlds' since they represent *possible worlds*.

What is crucial, but missing from the figure, is how Alice and Bob value the outcomes or worlds. Alice values the outcomes as

$$♡P_a ♣P_a ◊P_a ♠;$$

diamonds are not Alice's best friend, though not her worst. Bob's preferences are similar, but with one crucial difference

$$♠P_b ♣P_b ◊P_b ♡,$$

hearts and spades have completely swaped ends. An important feature of the PD is that *that is all the information that one has*; if any other information is added, it is no longer a PD. Now for the question: what should Alice and Bob do?

Rational choice says: do that which best satisfies your preferences. Since neither knows what the other is going to do, each should try and guarantee themselves the best payoff regardless of what the other does. That leads to each choosing D. In game theory (D_a, D_b) is called a Nash equilibrium which is defined as each player not doing better by making a different choice given that the other player holds its choice fixed.

But this is odd since if the players coordinated on (C_a, C_b) they would arrive at ♣ which both prefer (strictly) to ◊. The outcome that would seem to be the rational choice is not optimal. But that better situation is not stable. Once one player decides to do C, the other player would do better by doing D. (C_a, C_b) is not an equilibrium. It seems that the theory of rational choice suggests doing something that isn't optimal.

One of the lessons that Binmore (2007) suggests be learned from this toy example is not that there is something wrong with our theory of rationality, but that what the example provides is precisely the kind of situation where what we would usually call 'cooperation' cannot exist. Schotch sees things somewhat differently.

Schotch's issue is that the diagnosis of the situation in the prisoner's dilemma is unsatisfying. 'Why', we might hear him asking, 'is it that we see this as a dilemma at all given that we understand rationality—as game theory—and the reasoning involved is in accord with rationality?' His diagnosis of what goes wrong in the prisoner's dilemma is, roughly, that if one looks at society's preferences as revealed by the choices suggested by the standard theory of rational decision in the prisoner's dilemma, one would think society is crazy. How he gets there is by thinking about social choice.

11.3.1 Social Choice: Arrow vs. Schotch

Social choice theory is the mathematical theory of how to aggregate individual decisions into collective decisions. The early discussions of this from the mid 20th century focused on aggregating preferences, that is combining the preferences of individuals into a collective or societal preference relation.

An Arrovian Social Welfare Function, as it is called, is supposed to map a set of rankings into a single societal ranking of social states. Each function A takes a set of preference relations (one for each member of the society N) to a social preference ranking of W: $A(\{\,R_i : i \in N\,\})$, which I will denote R_N. It is assumed that the societal ranking is transitive and complete, but also must obey a set of seemingly benign conditions:

UD *Universal Domain* A must assign any collection of individual rankings a social ranking

ND *Non Dictatorship* no individual $i \in N$ should decide what the ranking of $x, y \in W$ should be regardless of the other inputs $R_j, j \in N$

IIA *Independence of Irrelevant Alternatives* The societal ranking of $x, y \in W$ shouldn't depend on the rankings of other alternatives in W

PP *(Strong) Pareto Principle* If everyone ranks x at least as good as y, and at least one individual i thinks xP_iy, then xP_Ny.

As is well known, these conditions are not all satisfiable. Indeed, Arrow's famous theorem shows that IIA, PP, and UD imply not ND. The standard social choice version of forming societal preferences is:

$$\text{Individual Preferences} \to^A \text{Societal Preference}$$

The idea seems to have been to conjure a general, impartial and rational way of satisfying preferences without getting one's hands dirty in politics. But, what Arrow's theorem showed was that no such process could be conjured. This is what Schotch sees as the paradox of social choice since we actually do make social choices often. The problem that I think Schotch saw with the social choice approach was that there was no politics involved; to work politics should be re-integrated into the mathematics.

Schotch sees societal preference as determined by a multistage process. First each member i of the society, N, must determine their individual preferences R_i. These preferences are used to form goals which reflect an individual's ideals for the overall social outcome, i.e., a set of social states they would like to achieve. That process is already a political process which may deform the individual's preferences. I will call that process P. These goals are then used to determine a societal goal by taking the intersection of all of the individual goals which come out of the political process. A social preference ordering can also be determined through a process which Schotch calls 'Type Lowering' (TL). Diagrammatically,

$$\text{Individual Preferences} \to^P \text{Individual Goals} \to^\cap \text{Societal Goal} \to^{TL} \text{Societal Preference}$$

11.3. SCHOTCH'S TAKE ON THE PRISONER'S DILEMMA

We see a tension in Schotch's version of social choice because the political process is where most of the "work" is done, but the forming of the individual goals isn't necessarily a purely rational process. Thus there isn't necessarily a uniform formal theory to be injected or formulated for that process. Nonetheless, Schotch has ideas about how individuals might formulate their goals that admits of a somewhat uniform formulation, but only requires the limited information of the individuals' preferences over states.

Consistent with traditional social choice theory for each $i \in N$, the following properties of preferences hold:

- For each $i \in N$ and $x, y \in W$ exactly one of xP_iy, yP_ix or xI_iy holds,
- P_i is transitive and irreflexive, while
- I_i is only reflexive.

These assumptions are at odds with those made in section 11.2. One then defines a relation R_i as

$$xR_iy \iff (xP_iy \text{ or } xI_iy),$$

which can be read as x is at least as good as y. Of course, xP_iy holds when xR_iy and not yR_ix, while xI_iy iff xR_iy and yR_ix. Since I_i isn't assumed to be transitive, R_i isn't transitive either, but it is complete, i.e., every pair of social states is inter-comparable by R_i. For simplicity I will assume I_i to be transitive.

Individual goals can be formed in a fairly straightforward manner: Choose the best elements from W according to R_i. The goal for i, according to this route will be

$$G_i = \{ x \in W : \forall y \in W, xR_iy \}.$$

But this can lead to problems. First, notice that $G_i = \max(W)$. As I have already said, if W is infinite, then this definition may only return an empty set in some case. Usually, social choice theory assumes W is finite. But there is another problem. The most natural way to generate a societal goal, then, is to take the intersection of the individual goals. But that can be problematic if there is no overlap between some of the individual goals. That situation is avoided in the PD example because although $\{\heartsuit\}$ may be the goal for Alice and $\{\spadesuit\}$ the goal for Bob, the only *feasible* goals are determined by the choices that are available to Alice and Bob.

But when given a set of feasible options as in the PD, another problem arises. The preferences that are given are those over individual social states, not sets of social states. What is required is a way to raise the preference relation from states to sets of states. Schotch suggests the following definition of such a type-raised relation $\leqslant$ between sets of states. For $X, Y \subseteq W$,

$$X \leqslant_i^S Y \iff [\forall y \in Y \exists x \in X \text{ s.t. } xR_iy] \text{ and } [\forall x \in X \exists y \in Y \text{ s.t. } xR_iy].$$

What this says is that for each state in Y there is some state in X which is at least as good as the one in Y, *and* for each state in X there is a state in Y which the one in X is at least as good as. Schotch sees this relation as being better than other relations offered in the literature on axiological versions of deontic logic, i.e., Jennings (1974) and van Fraassen (1973).

Aside on Type-raising

The scheme above can be interpreted as providing a generalization of the reasoning that is found in the PD. Although there are other schemes that would allow us to rank D_a above C_a, for example:

(J) $[\forall y \in Y \exists x \in X \text{ s.t. } xR_i y]$

(v) $[\exists x \in X \forall y \in Y \text{ s.t. } xR_i y]$.

But each of these definitions would rank D_a above C_a even if we changed the preferences. Suppose that Alice's preference profile now looked like:

$$\heartsuit P_a \clubsuit P_a \spadesuit P_a \diamondsuit$$

Then D_a would still be ranked as the highest option using either definition J or v, even though (D_a, D_b) is no longer a Nash equilibrium: v and J are not faithful representations of game theoretic reasoning. But Schotch's definition avoids *that* mismatch, however, Schotch's scheme doesn't guarantee a non-mismatch. Suppose Alice's preferences are:

$$\heartsuit P_a \spadesuit P_a \diamondsuit P_a \clubsuit,$$

then, since both of $\heartsuit$ and $\diamondsuit$ beat $\clubsuit$, $D_a \leqslant_a C_a$. But Alice would prefer to play C_a if Bob were to play D_b. Thus, (D_a, D_b) isn't a Nash equilibrium, even though D_a would be what is suggested to Alice according to Schotch's scheme. The Scheme that does get it right is:

$$X \leqslant_i^H Y \iff \forall x \in X, y \in Y, xR_i y$$

The H Scheme represents dominance reasoning. The disadvantage to using that scheme is that it is even more incomplete, generally speaking, than Schotch's. Indeed, if $X \leqslant_i^H Y$, then $X \leqslant_i^S Y$, but not necessarily the converse. So there is a trade-off between recommendations which result in Nash equilibrium and levels of completeness in comparability between sets at this abstract level. These problems can be set aside in what follows, all that is needed is a way of selecting goals when given a range of options or an agenda. **End of aside**.

Once a set of goals $\{ G_i : i \in N \}$ has been established, whatever "political" process is used to arrive at that, then a societal preference ordering can be determined. The intersection of the individual goals is referred to as the *societal goal*, and the following 'type-lowering scheme' is used to generate the societal preference relation:

$$xR_N y \iff [\forall i \in N, y \in G_i \Rightarrow x \in G_i].$$

In general R_N will not be a complete ordering. But we can order many of the elements. If $x, y \notin \cup \{ G_i : i \in N \}$, then $xI_N y$ since both $x \notin G_i$ and $y \notin G_i$ for all $i \in N$. If $x \in \cup \{ G_i : i \in N \}$, but $y \notin \cup \{ G_i : i \in N \}$, then $xP_N y$ since there is some G_i such that $x \in G_i$, but $y \notin G_i$, so not $yR_N x$. If $x, y \in \cap \{ G_i : i \in N \}$, then $xI_N y$, and if $x \in \cap \{ G_i : i \in N \}$ while $y \notin \cap \{ G_i : i \in N \}$, $xP_N y$: there will be at least one G_i such that $x \in G_i$ and $y \notin G_i$.

11.3. SCHOTCH'S TAKE ON THE PRISONER'S DILEMMA

The only case incomparability between states might arise is for those which are in

$$\cup\{G_i : i \in N\},$$

but not in

$$\cap\{G_i : i \in N\}.$$

They cannot in general be ordered based on this meager information. One needs more information about the set of goals. That will be the case whether or not $\cap\{G_i : i \in N\} \neq \varnothing$, i.e., whether or not the set of goals is consistent or compatible. The condition under which R_N is complete actually implies that there be coherence and quite a bit of consensus.

Observation 3. *R_N is complete iff $\{G_i : i \in N\}$ forms a chain.*[4]

Proof. Suppose R_N is complete. For reductio, assume $\{G_i : i \in N\}$ is not a chain, i.e., it is not linearly ordered by $\subseteq$. That means there are (at least) $j, k \in N$ such that $G_j \nsubseteq G_k$ and $G_k \nsubseteq G_j$. Thus, there are $x \in G_j$ and $y \in G_k$ such that $x \notin G_k$, $y \notin G_j$ and $x, y \notin \cap\{G_i : i \in N\}$. But that means neither $xR_N y$ nor $yR_N x$, contrary to completeness.

Conversely, assume $\{G_i : i \in N\}$ is linearly ordered by $\subseteq$. Suppose

$$x, y \in \cup\{G_i : i \in N\} \setminus \cap\{G_i : i \in N\},$$

which is the only case where states may not be related by R_N. Let G_k be the $\subseteq$-minimal set in $\{G_i : i \in N\}$ in which x appears, and similarly for y and G_j—there must be such sets since N is finite. Since $\{G_i : i \in N\}$ is a chain, either $G_k \subseteq G_j$ or vice versa. Assume $G_k \subseteq G_j$. If $y \in G_i$, then $G_i \supseteq G_j$, thus $G_k \subseteq G_i$, and so $x \in G_i$. Since G_i was arbitrarily chosen, $xR_N y$. Next assume $G_j \subseteq G_k$. If $x \in G_i$, then $G_i \supseteq G_k$, thus $G_j \subseteq G_i$, and so $y \in G_i$. Since G_i was arbitrarily chosen, $yR_N x$. Therefore, either way, at least one of $xR_N y$ or $yR_N x$. □

However, regardless of the level of consensus, when the goals are compatible, i.e.,

$$\cap\{G_i : i \in N\} \neq \varnothing,$$

Schotch has argued that goal formation using R_N will result in $\cap\{G_i : i \in N\}$. More precisely,

Theorem 1. *If $\cap\{G_i : i \in N\} \neq \varnothing$, then*

$$\cap\{G_i : i \in N\} = \{x \in W : \forall y \in W, xR_N y\}.$$

Proof. See Schotch (2016, p. 36) □

---

[4]This is why the sphere semantics from Lewis (1973) is extensionally equivalent to a relational semantics for total pre-orders.

What Schotch argues from there is that this process does what Arrow could not, although not how Arrow would have liked, see Arrow (1951). Schotch's method of deriving a societal preference order arises from individual preference orders but in an indirect way. Even though Schotch sees it as a way to get around Arrow's paradox of social choice, it doesn't—as Schotch points out—contradict Arrow's theorem. Schotch's method doesn't meet IIA or PP. It is well known that IIA requires disregarding information about the relative rankings of a pair x, y in an individual's ranking x, y. So if an individual's preference is $xPyPzPw$, then changing it to $zPwPxPy$ should have no effect on the societal preference of x with respect to y. But we can see that the composition of an individual's goal may be affected by such a change which can affect the societal ranking, contrary to IIA. PP can clearly fail, for suppose that every one ranks xPy, but that x, nor y are in any of the individual goals. Then $xI_N y$ since they are outside of $\cup \{ G_i : i \in N \}$, contrary to PP.

Non dictatorship and UD are a bit different. Schotch's process as has been described isn't really *one* process. It is a process schema. The schema must be filled out by augmenting it with a political process by which individual goals are determined. Universal domain simply requires that any combination of individual preferences be able to go through one of these processes. This is just a stipulation about what is going to count as an acceptable augmentation of the process schema. What might be more interesting is to look at what relationship must be maintained or present between the individual goals that are produced by the political augmentation of the schema and the individual preference orders which are input. As for non dictatorship, Schotch's method is and isn't dictatorial. Although the process schema could end up being dictatorial, all that must be done is require that none of those processes be allowed. However, there is a corollary to observation 3 above that shows complete rankings to be, sort of, dictatorial.

Corollary 1. *Let R_N be a complete Schotch ranking based on $\{ G_i : i \in N \}$. Since N is finite, there is $i \in N$ such that $G_N = \cap \{ G_i : i \in N \} = G_i$.*

Proof. Since $\{ G_j : j \in N \}$ is a chain by observation 3 and N is finite, there is a $\subseteq$-minimal element, call it, G_i. Then $G_i \subseteq G_j$ for all $j \in N$, so $G_i = \cap \{ G_j : j \in N \}$. □

Thus, the societal goal will be the the goal of some individual. This isn't dictatorship in Arrow's sense for two reasons. First, that one individual doesn't decide the rankings of all the states, that individual just decides what states are at the top of the ranking. Second, and most importantly, it may not be the same individual for any set of individual goals; it is just that at least one individual gets the top goal right. Put another way, there is at least one individual who is "in tune" with society's preferences. So much so that what s/he wants most, is what *everyone* wants most. That means completeness of social ordering requires consensus and coherence to a fairly high degree, at least for finite societies. Everyone's goals must choere with one individual's goal. My primary point is that Schotch's method, qua schema, is closer to what Arrow's conditions require than he seems to have thought, right down to non-dictatorship—in a certain loose sense—being at odds with completeness.

11.3.2 Back to the Dilemma

To diagnose the PD Schotch uses a particular augmentation of his schema. He doesn't implement any complicated political process for forming the individual goals. The process of individual goal formation is done by taking the highest ranked choice as determined by $\leqslant_i$. That is, after all, the process individual rationality suggests. In the PD above, the goals formed are D_a and D_b, as one would expect.

But now the process of societal goal formation is just $D_a \cap D_b$ which results in $\{\diamond\}$ as the societal goal. Applying the societal ranking formation one can see the following relationships. First, the ordering isn't complete since the individual goals do not form a chain by observation 3. Indeed, there is no relationship between ♠ and ♡. However, ◊ is strictly preferred to everything else. But one also gets that ♡ and ♠ are strictly preferred to ♣. So one gets a ranking that looks sort of like:

$$\diamond P_N \heartsuit ? \spadesuit P_N \clubsuit$$

Although Schotch frames his version with numbers, as is usually done, I have chosen a more abstract version to avoid committing cardinal sins. What Schotch says about this is as follows:

> We wouldn't rank [things in this way] unless we value defection or despise cooperation beyond all rationality. And *that* is the paradox: when we defect we are playing the same move that would be played by somebody else with crazy preferences. The two of us end up with the same goal after all. (Ibid., p. 39)

What Schotch means here is that if the societal preferences (which result from societal type lowering) are used to form another goal G_N, it is identical to the goals of each individual, i.e., $\{\diamond\}$, that follows from theorem 1. I think what Schotch is suggesting is that his process indicates what preferences are displayed when agents act interactively in the way that game theory suggests. But put in this abstract way, one can see two complications with his view of the situation.

First, a preference for defection or aversion to cooperation doesn't come into it. Strictly speaking, nothing about the states requires interpreting them as having anything to do with cooperation or defection. Second, the abstraction formulation doesn't represent anything as inherently odd about the type-lowered preference relation—other than that it is incomplete. What makes it odd is what made the PD look odd to begin with: the outcome is Pareto inefficient.

But Schotch's diagnosis is still correct, *if* it is interpreted somewhat differently. What is odd is that the societal preference isn't the preference of anyone in the society. Indeed, everyone strictly prefers ♠ to ◊, so the societal preference ranking is *at odds* with everyone's preferences. As a society engaged in competitive interaction, game theory requires everyone to collectively disregard a preference *everyone* has.

I think this observation can be summarized in an impossibility theorem for Schotch Social Choice Schemata. Say a a SSCS satisfies Universal Rational Domain when the individuals goals are determined simply by $\max(W)$ for each i in any decision situation. Also say a SSCS rule satisfies the Special Pareto Principle iff for all $x, y \in W$, if for all $i \in N$, xR_iy, and xP_jy for some j, then $y \in G_N$ only if $x \in G_N$.

Theorem 2. *There is no SSCS satisfying Universal Rational Domain and the Special Pareto Principle.*

Proof. The PD situation is included by Universal Rational Domain, but doesn't satisfy the Special Pareto Principle. □

If everyone has a certain preference, shouldn't the people, qua collective, respect that? Game theory says no in the case of the PD because the PD is about *competitive, individually motivated interaction under uncertainty*. If one attributes a partiality to the Pareto principle to Peter, then I think things can be seen his way: when we do as game theory suggests, we are not acting rationally qua collective. But ignoring such partiality towards collectives is what the PD is set up to do.

Thus, I remain with Binmore in saying that the PD is a situation in which the deck is stacked against cooperation. The reason why, however, can be highlighted using Schotch's method of deriving societal preferences. Individual rationality, represented by using rational domains, is at odds with collective interests as represented by the special Pareto Principle. However, other-regarding political processes are all around us all of the time, but since there is no other-regarding political process built into the PD, the outcome strikes us as odd. Schotch's method makes plain the disregard for others present in the PD by allowing us to see the damage it does to *everyone's* preferences. This is why it is the *prisoner's* dilemma and not the *prisoners'* dilemma since each prisoner is acting by themselves, not together. It is to that consideration of how collective obligations can be computed that I will turn now.

11.4 Type Raising in Deontic Logic

11.4.1 Prolegomenon to Deontic Logic

The version of deontic logic I will present here is based on an account of action inspired by game theory, and relates to the stit and xstit theories of action cf. Belnap et al. (2001) and Broersen (2014), respectively. Formally speaking, given a set of states of the world W, the performance of an action can be described abstractly via a function E between a triple consisting of some $s \in W$, $i \in N$, $a \in A_i$ (i's repertoire of actions at s), and some $X \subseteq W$. When $E(s, i, a) = X$, the action a by i in the state s has guaranteed that the world will be amongst the outcomes/states in X. A more compact way to describe this is as a set of functions $\{ E_i \}_{i \in N}$ each of which maps pairs (s, a) to subsets of W. A_i is really a function which provides the actions available to i at s, i.e., $A_i : W \to V$ where V is the set of all actions. I will stipulate that each agent must be able to perform each action $a \in A_i(s)$, and I will only write $a \in A_i$ leaving s implicit. Which outcome or particular state of W will result from i's action? That may not be certain. Given this uncertainty, the importance of type raising becomes apparent since preference relations are relations between states and not sets of states. Type raising is a way to

11.4. TYPE RAISING IN DEONTIC LOGIC

make decisions under complete uncertainty and is a way of comparing the results of actions given the uncertainty of their particular outcomes.

But there is another complication. In game theory, each group can be thought of as having its own repertoire consisting of combinations of the individual strategies of the group's members. To represent this in this general setting, each group also has a repertoire of actions $A_G = \{\,(g_i : i \in G) : \forall i \in G\, g_i \in A_i\,\}$ (at a particular s). There are two special cases of this: an action profile for the whole group N, $\bar{n}$, and the action profile for the empty group $\bar{o}$. The latter consists of no actions at all, so it is unique.

In game theory, unlike in stit theory, the effect of a group G's action can be different from the combined effect of the members' actions $\bar{g} = (g_i : g_i \in A_i)$. Put formally, I will not define $E_G(s, \bar{g})$ as, $\cap\{\,E_i(s,g) : (g \in \bar{g}, g \in A_i, \,\&\, i \in G)\,\}$ where $\bar{g}$ is an action profile consisting of one action from each member of $G \subseteq N$. Instead, I will treat the effect of an action profile $\bar{g}$ as a subset of $\cap\{\,E_i(s,g) : (g \in \bar{g}, g \in A_i, \,\&\, i \in G)\,\}$. The following are additional properties which relate to properties of outcome functions of game forms:

(a) for all s, $G \subseteq N$, and $\bar{g} \in A_G$, $E_G(s, \bar{g}) \neq \emptyset$ (that includes $G = \emptyset$),

(b) for each $\bar{n} \in A_N$ at s, there is $w \in W$ such that $E_N(s, \bar{n}) = \{\,w\,\}$,

(b') for all s and $w \in E_\emptyset(s, \bar{o})$, there is $\bar{n} \in A_N$ such that $E_N(s, \bar{n}) = \{\,w\,\}$,

(c) for all $G \subseteq B \subseteq N$, if $\bar{g} \subseteq \bar{b}$, then $E_B(s, \bar{b}) \subseteq E_G(s, \bar{g})$,

(c') for all $G \subseteq N$, $\bar{g} \in A_G$, $E_G(s, \bar{g}) = \cup\{\,E_N(s, \bar{g}, \bar{a}) : \bar{a} \in A_{N \setminus G}\,\}$, and

(d) if $G \cap B = \emptyset$, then for all $\bar{g}$ and $\bar{b}$, $E_G(s, \bar{g}) \cap E_B(s, \bar{b}) \neq \emptyset$.

A few comments are in order. Imposing all of these conditions allows one to define a function $E : W \times \mathcal{P}(N) \to \mathcal{PP}(W)$ such that $E(s, G) = \{\,E_G(s, \bar{g}) : \bar{g} \in A_G\,\}$. Defined as such, each $E(s,\ ) : \mathcal{P}(N) \to \mathcal{P}(\mathcal{P}(W))$ is an effectivity function. Indeed, the conditions a-d make the function a *truly playable* effectivity function, cf. Goranko et al. (2013). And those effectivity functions correspond to the α-effectivity function of some game form, cf. Abdou and Keiding (1991). In the literature on effectivity functions condition (b) and (b') together are referred to as citizen sovereignty—the actions of all the agents completely determine the particular outcome and they can determine every possible outcome. (c) is called coalition monotonicity; the larger the group the more refined the outcome it can ensure. (c') would make the outcome of a group's action be analyzable as the union of all the possible effects that could result when the agents outside the group act as well. Such a property makes sense in the context of a game form because outcome functions in game forms are only defined for the complete strategy profiles. Thus $E_G(s, \bar{g})$ would have to be reverse engineered from the $E_N(a, \bar{g}, \bar{a})$. Obviously, condition (c') entails condition (c). (d) is a stronger version of the property known as regularity which is a consistency condition—disjoint groups can't block things from happening altogether. Together conditions (a), (c) and (d) mean that the effectivity function is superadditive. Whether one wants to impose all of these conditions depends on how much like a game form one wants the model to be. For my part, I will take conditions (a), (c) and (d) as gospel. That will make applications for stit and xstit theory easier.

But the question is: given that multiple actions are available to an agent, or group of agents, which should the agent choose to do? Which set of outcomes should or ought an agent pick? The response is simple from an axiological standpoint: the best one. Then one asks: how is the best one determined?

Above I introduced preference relations R_i on states in W, and offered different ways to use that preference relation to provide a preference relation on sets of those items. If W is interpret, not as items, but as social states or states of the world, then a preference relation between states can be used to generate a preference relation between sets of states by type raising.

So what one ought to do, is that action with the best set of outcomes. But again there is a problem. Even if a linear ordering is definable by $\preccurlyeq^*$ it may not follow that there is a best set. At a state s there may be infinitely many actions available, say $X_1, X_2, \ldots X_k$ for $k \in \mathbb{N}$. Each outcome is better than the previous, $X_i \prec^* X_j$ for $i, j \in \mathbb{N}$. In this case there is no best action. Provided that an ordering is total and there are only a finite set of elements to consider, then there will be a best set. There is no guarantee, however, that $\preccurlyeq^*$ generates a total ordering, even when the set of sets to compare is finite. Indeed, some sets are incomparable.

But a set of *optimal* outcomes can be defined relative to a relation $\preccurlyeq^*$ on $\mathcal{P}(W)$ as follows:

Definition 3. A set $X \subseteq W$ is optimal with respect to $\preccurlyeq^*$ iff there is no $Y \subseteq W$, $X \prec^* Y$, i.e., there is no set in $\mathcal{P}(W)$ which dominates X. We denote the set of optimal sets from among a set of sets $X \subseteq \mathcal{P}(W)$ as $Opt(X)$. I.e., $Opt(X) = \{ Y \in X : \text{not } \exists Z \in X \text{ s.t. } Y \prec^* Z \}$.

If the set of available actions, i.e., A_i, is always finite, then the set of correlated outcomes will be finite. That is, since each A_i is finite, each A_G will be finite for $G \subseteq N$ (N is also finite), and then $\{ E_G(s, \bar{g}) : \bar{g} \in A_G \}$ will be finite. I will refer to this last set as $E_G(s)$. Thus, when each A_i is finite, $Opt(E_G(s))$ will be non-empty.

Even if there are optimal outcomes, there may not be a *best* outcome. Thus, there may not be any *best* action. But there will always be at least a selection of acceptable actions: those that correspond to the optimal outcomes.

There is still a gap: $\preccurlyeq^*$ is supposed to be derived from a preference relation between states. But the preference relations are those of the individuals in N, while it is the actions of groups that are under consideration. So there are, at least, a couple of possibilities present here: (1) should there be *one* preference relation which determines Opt somehow derived from the preference relations of all the individuals? or (2) should each group determine a preference relation for its members and use that to determine what is optimal for them to do?

What would suit the usual conception of practical reasoning would be neither (1) nor (2), but a non-other regarding multi-stage procedure where one looks at how one can guarantee *themselves* a good outcome regardless of what the others do, or trying to figure out what action guarantees others the best outcome regardless of what everyone

11.4. TYPE RAISING IN DEONTIC LOGIC

else does, and then using that as the basis for reasoning about what one should do. That is the kind of reasoning deployed in determining a Nash equilibrium and what leads to the (D_a, D_b) outcome in the prisoner's dilemma. In works like Horty (2001), an objective preference relation, i.e., one not relative to individual agents, is used to determine what to do. I will not take a stand on which of these is *the* right way to look at the matter. Rather I will follow Tamminga (2013) in allowing groups to use the interests of other groups in determining what to do.

So let's take a moment to summarize the formal model so far. Each $s \in W$ has assigned to it a set of actions for each agent i, A_i. There is then a set of effects of action profiles determined by the $\bar{g} \in A_G$ and $G \subseteq N$, $E_G(s)$. The set of effects is then ordered according to $\preccurlyeq^*$, and there is a set of optimal actions $Opt(E_G(s))$. $Opt(E_G(s))$ is non-empty when $|E_G(s)|$ is finite. But at the moment I am leaving how $\preccurlyeq^*$ is determined open. This model is to provide an interpretation for a language defined as follows:

$$\varphi := \mathbf{p} \mid \neg\varphi \mid O(a) \mid P(a) \mid O\varphi \mid P\varphi \mid \varphi \wedge \varphi \mid \varphi \supset \varphi$$

p is an atomic sentence from a set of atoms **At**. O has two parameter spots that can be filled in with groups as follows: $O_{G,B}$. $O_{G,B}(a)$ is then read as 'B ought to do a in G's interest' where a is an action in B's repertoire. $O\varphi$ is similar can can be filled out as $O_{G,B} \varphi$. $O_{G,B} \varphi$ is read as 'B ought to make it so that φ according to G's interests'; mutatis mutandis for P using 'permission' and its cognates. Both conceptions of ought are versions of ought to do as opposed to ought to be. Later I will deal with an operator $O_G \varphi$ that expresses an ought to be impersonal obligation. But both of these versions of ought to do will have an axiological basis rather than a purely deontic grounding.

Definition 4. An *Axiological Frame* $\mathfrak{F} = \langle W, N, \{A_i\}_{i \in N}, \{R_i\}_{i \in N}, \{E_i\}_{i \in N}\rangle$ is a quintuple consisting of a non-empty set W; a finite set of agents N; a set of action repertoires for each agent; a set of preference relations, one for each agent; and a function E_i for each agent as described above. If for each $s \in W$, and $G \subseteq N$, A_G is finite relative to s, then it is a *finite choice* axiological frame.

An axiological model $\mathfrak{M}$ is an axiological frame $\mathfrak{F}$ along with a function $v : \mathbf{At} \to \mathcal{P}(S)$. The topic of the following sections will be ways to explicate the various truth conditions for obligation and permission operators.

11.4.2 Semantics for Deontic Terms

In this paper I am giving axiological accounts of the deontic terms; I equate that with a roughly consequentialist stance on the evaluation of action. Thus, each way of providing that semantics makes reference to some way of assigning value to the effects of actions: $\preccurlyeq^*$. If Tamminga (2013) is followed, any group can follow the values of any other group in deciding how to act. One can define $\preccurlyeq^*_G$ via type raising R_G—the preference relation for the group G. How R_G is determined or whether it can be determined is left open. It is also left open how to type raise R_G. I follow Tamminga again in focusing on the particular terms of 'permission' and 'obligation'.

The permitted acts are those actions whose effects are optimal and the obligatory acts are those actions whose effects are at least as good as any available act. I will refer to this pair of deontic terms as *deontics*. In what follows, I will simply peak of acts are optimal (better) rather than correctly referring to their effects. A group G has an *other regarding* conception of deontics when the notion of value G uses in deciding what to do is derived from another group. It is *self regarding* otherwise.

Amongst the self and other regarding values that are used, there are also different ways in which one can think about how what is optimal is affected by the actions of others. I may know that there are 25000 people needing food in the north and 5000 in the west. And I have sufficient trucks to send food west and north. But I may also know that if I send trucks west now, they will be ambushed by bandits before they deliver the food, and I know that the bandits will be gone if I wait a few days. But I know that some of those 5000 will die, if I wait. Thus, I decide that, in the interest of the majority, I only send trucks north *now*. The point is that I act in a way where what the best options are is constrained by what other groups of agents might do. This is the standard reasoning that takes place in game theory. The general concept of comparison that is used is one where we rank the actions we choose by what actions others might take. This leads to ranking the action profiles themselves via their effects when others have also acted. Formally, we have the following definition.

Definition 5 (Action Ranking). Given a state s, a group B who is acting, $\leqslant_G^*$ based on the interests of a group G, and action profiles $\bar{b}$ and $\bar{b}'$ in $A_B(s)$,

$$\bar{b}' \leqslant_{s,B,G} \bar{b}$$

$$iff$$

$$\forall \bar{a} \in A_{N \smallsetminus B} \left[E_N(s,(\bar{a},\bar{b}')) = E_N(s,(\bar{a},\bar{b})) \text{ or } E_N(s,(\bar{a},\bar{b}')) \leqslant^* E_N(s,(\bar{a},\bar{b})) \right]$$

Informally, $\bar{b}$ is at least as good as $\bar{b}'$ for G relative to $\leqslant^*$ when for any action $N \smallsetminus B$ might take, call it $\bar{a}$, either the outcomes of $\bar{b}$ with $\bar{a}$ and $\bar{b}'$ with $\bar{a}$ are identical, or the outcome of $\bar{b}$ with $\bar{a}$ is better than that of $\bar{b}'$ with $\bar{a}$. If condition (b) on the effectivity function E_N was used, a typed raised relation isn't really necessary; R_G could be used to compare the outcomes of actions. But if one relaxes (b), as I have, and just keep (b') or not, the definition above is what is needed. I think that indeterminacy is fundamental. So I assume that we cannot use R_G to compare the outcomes of actions. We have to use a type raised relation based on R_G.

We can of course define a strict version of the relation on actions as follows:

$$\bar{b}' <_{s,G,B} \bar{b} \iff [\bar{b}' \leqslant_{s,G,B} \bar{b} \ \& \ \bar{b} \not\leqslant_{s,G,B} \bar{b}']$$

Here, then, we have our first conception of the deontics. An action (profile) is permitted for a group G iff it is not strictly worse than any other action (profile) for the group relative to $\leqslant_{s,G,B}$. An action (profile) is obligatory for a group G iff it is at least as good as the other available actions. Note that that means it must be comparable to all of the other actions. These operators include a number of other parameters just like $\leqslant_{s,G,B}$ does. What we have is a semantics for $O_{G,B}(a)$ and $P_{G,B}(a)$. Put formally,

- $\mathfrak{M}, s \Vdash O_{G,B}(\bar{b})$ iff for all $\bar{b}' \in A_B$ at s, $\bar{b}' \leqslant_{s,G,B} \bar{b}$

11.4. TYPE RAISING IN DEONTIC LOGIC

- $\mathfrak{M}, s \Vdash P_{G,B}(\bar{b})$ iff for all $\bar{b}' \in A_B$ at s, $\bar{b} \not<_{s,G,B} \bar{b}'$

I will keep with tradition and say that prohibition is when the action isn't permitted:

- $\mathfrak{M}, s \Vdash F_{G,B}(\bar{b})$ iff there is $\bar{b}' \in A_B$ at s such that $\bar{b} <_{s,G,B} \bar{b}'$

That is to say, an action is forbidden when it is worse than another action. Before moving on, I want to note a few things. It is possible to be obligated to do more than one action. And it is possible for the effects of those obligations be inconsistent. But if an action is obligatory, it is permitted. Apart from these operators there is also the 'ought to bring about' operator: $O_{G,B}\, \varphi$.

For that operator there are two definitions I want to consider. Consider Horty's definition from Horty (2001). In full generality, Horty's condition is:

- $\mathfrak{M}, s \Vdash O_{G,B}\, \varphi$ iff for all $X \in E_B(s)$ such that there is $s' \in A$ such that $\mathfrak{M}, s \not\Vdash \varphi$, there is $Y \in E_B(s)$ such that

(1) $X <^*_{s,G} Y$

(2) for all $s' \in Y$, $\mathfrak{M}, s' \Vdash \varphi$, and

(3) for each $Y' \in E_B(s)$ such that $Y \leq^*_{s,G,B} Y'$, for all $s' \in Y'$, $\mathfrak{M}, s' \Vdash \varphi$.

This truth condition says, for any action that doesn't ensure the truth of φ, there is an action that is strictly preferred that does, and any action at least as good as that one will also ensure the truth of φ. This truth condition handles infinite cases. Suppose that $\mathbf{q}$ is true for each X_i when $i > 3$. Then $O\, \mathbf{q}$ will be true. This is because for each of $A_1, \ldots, A_3$ there is another action, i.e., X_4, such that $X_j <^* X_4$, $\mathfrak{M}, s' \Vdash \mathbf{q}$ for all $s' \in X_4$, and for each $k \geq 4$, $\mathfrak{M}, s' \Vdash \mathbf{q}$. As Horty notes, however, when $Opt_G(E_B(s))$—the optimal effects in $E_B(s)$ relative to G—is non-empty, the previous truth condition is equivalent to:

- $\mathfrak{M}, s \Vdash O^1_{G,B}\, \varphi$ iff for all $Y \in Opt_G(E_B(s))$, if $s' \in Y$, then $\mathfrak{M}, s' \Vdash \varphi$.

That would mean $O^1_{G,B}\, \varphi$ is true when each optimal outcome that B is capable of bringing about ensures the truth of φ. So for finite choice axiological frames the two truth conditions match. If $[\![\varphi]\!]$ is the set of states that make φ true, then the truth condition can be rewritten as:

$$\bigcup Opt_G(E_B(s)) \subseteq [\![\varphi]\!].$$

But O^1 may not do the right work in a multi-agent setting since it doesn't consider how the effects of other's actions might influence the effect. But that can be imposed fairly easily. What one can do is consider only the propositions which are guaranteed by the permitted actions, that is actions whose effects are not dominated when the effects of others' actions is added. It can be defined as follows:

- $\mathfrak{M}, s \Vdash O^{1a}_{G,B}\, \varphi$ iff for all $\bar{b} \in A_B$, if $\forall \bar{b}' \in A_B, \bar{b} \not<_{s,G,B} \bar{b}'$, then $E_B(s, \bar{b}) \subseteq [\![\varphi]\!]$.

What ought to be ensured are the effects of those actions which are optimal according to G from the effects that B can have. This meshes with the concept of relative comparability above since it incorporates what the other agents outside of B might do. It also follows that $O_{G,B}^1 \varphi$ would imply $O_{G,B}^{1a} \varphi$. The conditions for permissibility then are just the existential versions of the conditions above. Since it needn't be that $E_B(s,\bar{b}) = \bigcup \{ E_B(s,\bar{b}\bar{a}) : \bar{a} \in A_{N \setminus B} \}$, there is another set of possible definitions.

- $\mathfrak{M}, s \Vdash O_{G,B}^2 \varphi$ iff for all $E_B(s,\bar{b}) \in Opt_G(E_B(s))$, for each $\bar{a} \in A_{N \setminus B}$, $E_B(s,\bar{b},\bar{a}) \subseteq \llbracket \varphi \rrbracket$; and

- $\mathfrak{M}, s \Vdash O_{G,B}^{2a} \varphi$ iff for all $\bar{b} \in A_B$, if $\forall \bar{b}' \in A_B, \bar{b} \not<_{s,G,B} \bar{b}'$, then for each $\bar{a} \in A_{N \setminus B}, E_B(s,\bar{b},\bar{a}) \subseteq \llbracket \varphi \rrbracket$.

So there is another way to interpret the definition of $O_{G,B} \varphi$. Of course, $O_{G,B}^{1a} \varphi$ implies $O_{G,B}^2 \varphi$, and $O_{G,B}^2 \varphi$ implies $O_{G,B}^{2a} \varphi$. For simplicity, I will suppose condition (c'). In that case, $O_{G,B}^2 \varphi$ is equivalent to $O_{G,B}^1 \varphi$ similarly for the 'a' versions. I will also take $O_{G,B} \varphi$ to be the weakest of the definitions under the assumption of (c'): $O_{G,B}^2 \varphi$. For clarity, the definition of $O_{G,B} \varphi$ is:

- $\mathfrak{M}, s \Vdash O_{G,B} \varphi$ iff for all $\bar{b} \in A_B$, if $\forall \bar{b}' \in A_B, \bar{b} \not<_{s,G,B} \bar{b}'$, then

$$\bigcup \{ E_N(a,\bar{b},\bar{a}) : \bar{a} \in A_{N \setminus B} \} \subseteq \llbracket \varphi \rrbracket,$$

which under the assumption of (c') the truth condition amounts to

- for all $\bar{b} \in A_B$, if $\forall \bar{b}' \in A_B, \bar{b} \not<_{s,G,B} \bar{b}'$, then $E_B(s,\bar{b}) \subseteq \llbracket \varphi \rrbracket$.

Thus, the obligatory states of affairs are those which are the case relative to every permissible action (according to G's interests). Immediately, there may be no obligatory actions in a situation, but nonetheless, there can be propositions G is obligated to bring about. This makes intuitive sense of something like 'I ought to see to it that I give to charity', but at the same time there is no particular charity, i.e., no particular action that I ought to do. The correlated permission operator is also a 'permitted to ensure' operator. So although it has an existential condition, it is not the dual of the obligation operator, in the same way that $P_{G,B}(\bar{b})$ and $O_{G,B}(\bar{b})$ are not dual.

- $\mathfrak{M}, s \Vdash P_{G,B} \varphi$ iff there is $\bar{b} \in A_B$, such that $\forall \bar{b}' \in A_B, \bar{b} \not<_{s,G,B} \bar{b}'$, and $E_B(s,\bar{b}) \subseteq \llbracket \varphi \rrbracket$.

The prohibition operator, if I am to keep with tradition, must then be:

- $\mathfrak{M}, s \Vdash F_{G,B} \varphi$ iff for all $\bar{b} \in A_B$, if $\forall \bar{b}' \in A_B, \bar{b} \not<_{s,G,B} \bar{b}'$, then $E_B(s,\bar{b}) \cap \llbracket \neg \varphi \rrbracket \neq \emptyset$.

There is another 'ought to be' truth condition in the literature, going at least back to van Fraassen (1973) and is used by Schotch in his formulation of goal theory Schotch (2001) and Schotch (2016), which can be interpreted in this framework as well. It interprets what propositions are obligatory by looking at what propositions are strictly

11.4. TYPE RAISING IN DEONTIC LOGIC

better than their negations. The rationale being something like: is it better for it to be φ or not to be φ? If one is strictly better, that is what should be. Formally, the truth condition was

- $\mathfrak{M}, s \Vdash O\varphi$ iff $[\![\varphi]\!] >^* [\![\neg\varphi]\!]$.

But that condition doesn't really make sense in the current framework; for one, it is not relative to any particular group's interests. It is also independent of the state of evaluation, φ must be universally better than $\neg\varphi$. The first step is to remove that universality. Since $E_\varnothing(s)$ just consists of one set of worlds, and every other effect of an action is amongst those worlds, I will use $E_\varnothing(s)$ to refer to the range of possible outcomes for action relative to s. This relative obligation can then be interpreted as:

- $\mathfrak{M}, s \Vdash O_G \varphi$ iff $[\![\varphi]\!] \cap E_\varnothing(s) >^*_G [\![\neg\varphi]\!] \cap E_\varnothing(s)$.

But this doesn't connect up with the reading of $O_{G,B}\varphi$ where it is supposed to be B which is under the obligation to ensure that φ. Such a requirement can be related to the thesis from Belnap et al. (2001) which says that the complement of an ought sentence which is saying that something is obligatory, rather than in the predictive sense of 'ought', should be a stit sentence. That requirement is, however, suspect. After all, people recognize moral ideals: non-world hunger is better than world hunger; the Syrian refugees having food and shelter is better than them not. I find that the 'ought to be' sense of 'obligation' is a means of expressing these normative ideals, particularly moral ideals.

In the next two sections I will deal with two topics relating deontics to type raising. The first looks at the effect of the type raising scheme on the relationship between the *ought to be* and *ought to do* operators. The second considers the effects of the type raising scheme on what a group ought to do verses what its subgroups ought to do.

11.4.3 Ought to Be and Ought to Do

Given this common framework for ought to be and ought to do one has the possibility of studying the relationship between the two concepts in a fairly direct manner. On that topic, I want to ask the following question: do normative ideals influence ought to do deontics? What this means within the framework is whether the existence of an ideal, i.e., the truth of $O_G \varphi$, implies anything for the $O_{G,B}$ operators? That is a rather broad topic so I will narrow it down a bit.

There are two questions that have fairly easy answers. First, if φ is an ideal, does that mean that someone/group has an obligation to bring φ about? In the case of the Syrian refugees, for example, sure it ought to be that they have food and shelter, but does anyone actually have an obligation to bring that about? Intuitively, I think that someone must have that obligation; indeed, many say the European Mediterranean countries have such an obligation.

Second, if φ is an ideal, and one can bring φ about, is it their obligation to do so? Suppose Italy can see to it that the refugees are cared for. Does that mean that they have an obligation to do so? The answer to both questions can be negative on this semantics.

To answer the first question it is best to answer the second question first. I have said that the answer is negative, but there is a related question which is positive which will shed some light on the second question. The question is: can someone be obligated to do something incompatible with an ideal?

My intuitions pull in both directions on that question. I can have conflicting obligations which arise from different sources of obligation. I can be obligated to help at my child's school, while also at the same time be obligated to meet a deadline at work, but that isn't the sort of obligation that is at issue here. Can we, when only using one concept or source of value, recognize an ideal, but have an obligation to bring about the negation of the ideal? The answer to that latter question is: it depends. And what it depends on is the type raising scheme.

To make the notation a bit more manageable let $[\![\varphi]\!]_s = [\![\varphi]\!] \cap E_\varnothing(s)$. Consider the following lemma.

Lemma 1. *Let $\mathfrak{M}$ be an axiological model, $G \subseteq N$, and $Y, X \subseteq W$. Suppose that $X' \subseteq X$ and $Y' \subseteq Y$.*

(1) If $Y <_G^H X$, then $X' \not<_G^H Y'$.

(2) If $Y <_G^J X$, then $X \not<_G^J Y'$.

(3) If $Y <_G^v X$, then $X \not<_G^v Y'$.

Proof. For 1.1, suppose $Y <_G^H X$, and for reductio assume $X' <_G^H Y'$. By the definition of $<_G^H$, all of the items in Y' are at least as good as all those in X', and there must be at least one $y \in Y'$ that is better than some item in X'. But that means there is some item in X that is strictly worse than some item in Y, but by definition $Y <_G^H X$ means all of the items in X must be at least as good as all of the items in Y. A contradiction.

For 1.2, suppose $Y <_G^J X$, and for reductio assume $X <_G^J Y'$. By the definition of $<_G^J$, for each item in X there is an item in Y' that is at least as good as the item in X, and there is an item in Y' that is strictly better than every item in X. Call that item y. But that means there is an item in Y that is strictly better than every item in X. By definition, $Y <_G^J X$ means that for each of the items in Y there is an item in X that is at least as good as the item in Y. But it is impossible for an item in X to be at least as good as y. A contradiction.

For 1.3, suppose $Y <_G^v X$, and for reductio assume $X <_G^v Y'$. By observation 1, there is an item in Y' that is better than all of the items in X (call it y). Also by observation 1, $Y <_G^v X$ means there is something in X strictly better than every item in Y, including y. A contradiction. $\square$

In a particular situation in an axiological model lemma 1 has an important implication. Suppose one is looking at what actions B has available at a state s. Also suppose that $[\![\varphi]\!]_s >_G^* [\![\neg\varphi]\!]_s$, i.e., φ expresses an ideal in G's interest. Suppose $*$ is H, that is the type raising scheme is H. Now also suppose that B can ensure that φ, i.e., there is $\bar{b} \in A_B$ such that $E_B(s, \bar{b}) = X''$ and $X'' \subseteq [\![\varphi]\!]$. So B can ensure that the ideal is realized. It is then impossible for B to be obligated to bring about $\neg\varphi$.

11.4. TYPE RAISING IN DEONTIC LOGIC

Observation 4. *Let $\leqslant_G^* = \leqslant_G^H$. Suppose, φ is an ideal in G's interest at s, and B can ensure φ at s, then $\mathfrak{M}, s \Vdash O_{G,B} \neg\varphi$.*

Proof. Let, $E_B(s, \bar{b}'') \subseteq [\![\varphi]\!]$, and suppose for reductio that $\mathfrak{M}, s \Vdash O_{G,B} \neg\varphi$. That means for any $\bar{b}$ that isn't worse than any other action, $E_B(s, \bar{b}) \subseteq [\![\neg\varphi]\!]$. Since $E_B(s, \bar{b}'') \subseteq [\![\varphi]\!]$, there must be $\bar{b}'$ such that $\bar{b}'' <_{s,G,B} \bar{b}'$. Hence, there must be $\bar{a} \in A_{N \setminus B}$ such that $E_N(s, \bar{b}'', \bar{a}) <_{G,B}^H E_N(s, \bar{b}', \bar{a})$. Since, $E_N(s, \bar{b}'', \bar{a}) \subseteq E_B(s, \bar{b}'') \subseteq [\![\varphi]\!]_s$ by coalition monotonicity, and similarly for $E_N(s, \bar{b}', \bar{a})$, by lemma 1.1 we have a contradiction. □

What that means is that if a group/agent can bring about an ideal, they cannot be obligated to do something inconsistent with it. That does not mean that they are prohibited from bringing about something inconsistent with it. If one of the optimal actions guarantees something inconsistent with the ideal, they are indeed permitted to bring about a non-ideal state. But they cannot be required to bring it about.

The case of using the J type raising scheme leads to a similar result, but for different reasons. It requires making stronger assumptions and using more properties of the effectivity function. In particular, the ability of the group must be stronger. What the group B must be capable of is that it can guarantee $[\![\varphi]\!]_s$ itself, not just a subset of it. Formally, that means $[\![\varphi]\!]_s \in E_B(s)$. If that is the case, then $\neg\varphi$ cannot be a proposition the agents ought to bring about—in G's interests.

Observation 5. *Let $\leqslant_G^* = \leqslant_G^J$. Suppose, φ is an ideal in G's interest at s, and B can ensure exactly $[\![\varphi]\!]_s$ at s, then $\mathfrak{M}, s \Vdash O_{G,B} \neg\varphi$.*

Proof. Let, $E_B(s, \bar{b}'') = [\![\varphi]\!]$, and suppose for reductio that $\mathfrak{M}, s \Vdash O_{G,B} \neg\varphi$. That means for any $\bar{b}$ that isn't worse than any other action, $E_B(s, \bar{b}) \subseteq [\![\neg\varphi]\!]$. Since $E_B(s, \bar{b}'') \subseteq [\![\varphi]\!]$, there must be $\bar{b}'$ such that $\bar{b}'' <_{s,G,B} \bar{b}'$ otherwise $E_B(s, \bar{b}'') \subseteq [\![\neg\varphi]\!]$. That means for every $\bar{a} \in A_{N \setminus B}$, $E_N(s, \bar{b}'', \bar{a}) \leqslant_{G,B}^J E_N(s, \bar{b}', \bar{a})$. Since $[\![\neg\varphi]\!] <_{G,B}^J [\![\varphi]\!]_s$, there must be $x \in [\![\varphi]\!]_s$ that is better than any element of $[\![\neg\varphi]\!]_s$ (see the comments in section 11.2). By property (c') of the function E,

$$[\![\varphi]\!]_s = E(s, \bar{b}'') = \bigcup \{ E_N(s, \bar{b}'', \bar{a}) : \bar{a} \in A_{N \setminus B} \}.$$

So for some $\bar{a}' \in A_{N \setminus B}$, $x \in E_N(s, \bar{b}'', \bar{a}')$. But then x is better than all of the elements of $E_N(s, \bar{b}', \bar{a}')$, thus $E_N(s, \bar{b}'', \bar{a}) \not\leqslant_{G,B}^J E_N(s, \bar{b}', \bar{a})$. A contradiction. □

The reasoning in the $\leqslant_G^* = \leqslant_G^V$ case is the same as in the J case. The analog of lemma 1 is not provable for the S type raising scheme. I will leave finding a counterexample as an exercise, but I will give a hint. For the S scheme, when $Y' >_G^S X$, there are two conditions that might fail for $Y' \leqslant_G^S X$. Either the J half of the condition will fail or the Onto half of the condition. Hence, the Onto half of the condition failing is compatible with $X >_G^S Y$. The reasoning in observation 5 isn't open in the S case either. Again the Onto condition is problematic. When it fails making $X <^S Y$ true, there must be something in X that is worse than every element of Y. But that worse element can be anywhere; it doesn't have to be in any of the $E_N(s, \bar{b}', \bar{a})$s.

The counterexample for the S case is somewhat complicated because so many things need to be specified.

- $W = \{s, x_1, x_2, y_1, y_2\}$, $Y = \{y_1, y_2\}$ and $X = \{x_1, x_2\}$
- The preference relation: $x_1 I y_1$, $x_1 P y_2$, $y_1 P y_2$, $(y_i P x_2 : i = 1, 2)$, $x_1 P x_2$ and everything in $X \cup Y$ is preferred to s,
- $N = \{1, 2\}$, $A_1(s) = \{a_1, a_2, a_3\}$, and $A_2(s) = \{b_1\}$,
- $E_1(s, a_1) = Y$, $E_1(s, a_2) = \{x_1\}$ and $E_1(s, a_3) = \{x_2\}$,
- $E_2(s, b_1) = X \cup Y = E_\varnothing(s, \bar{o})$, and
- $E_N(s, a_i, b_1) = E_1(s, a_i) \cap E_2(s, b_1)$ for $i = 1, 2, 3$.
- Finally, $v(\mathbf{p}) = Y$ and so $[\![\neg \mathbf{p}]\!] = X \cup \{s\}$.

As the reader can check, $[\![\neg\mathbf{p}]\!]_s <^S [\![\mathbf{p}]\!]_s$ since x_2 is worse than everything in $[\![\mathbf{p}]\!]_s$ and for any x there is a y that is at least as good. But, $Y <^S \{x_1\}$ and $\{x_2\} <^S Y$, also. Since a_2 is the only undominated action for 1, only a_2 is permitted, and $E_1(s, a_2) \subseteq [\![\neg\mathbf{p}]\!]$. Thus, $\mathfrak{M}, s \Vdash O_{G,B} \neg \mathbf{p}$. But 1 can ensure $Y = [\![\mathbf{p}]\!]_s$ exactly. So that leaves the possibility that one can be obligated to bring about propositions which are completely at odds with ideals, ideals that arise from the same considerations of value.

I will end this section with the kind of situation that I think makes intuitive sense of this. If one is in a bad situation, certain things become, not only permissible, but obligatory. These things are things that one would look at as being bad under normal circumstances. Here one can picture the situation in the Warsaw ghettoes. People may not have seen following their ideals as even being permissible in such circumstances, and it is metaethically possible, I think, that they were right. That is not to say that anything goes; there are certain things which are, perhaps, beyond the pale.

Schotch suggests just such a way to ensure that certain things are beyond the pale. He suggests that not just any proposition that is strictly better than its negation is acceptable. It should also be "before the pale". He introduces the idea of an *anti-goal*. An anti-goal is simply some proposition that is deemed beyond the pale. Thus, if φ is strictly better than $\neg\varphi$, it will be an ideal only if it is disjoint from all of the anti-goals. Such a solution introduces non-axiological considerations back into the deontic semantics, unless the anti-goals can be determined via some axiological means. For example, referring back to section 11.3.1, one might consider enough, say a majority, of individuals excluding something from their goals is sufficient to deem it beyond the pale. For example. Suppose Jeff wants to exploit children for economic gain. Thus, it is one of Jeff's goals for Jeff to exploit children. Let's call that proposition **JEC**. That proposition is contained in the 'exploit children' proposition, call it **EC**. Suppose that a majority in Jeff's society are such that $G_i \cap [\![\mathbf{EC}]\!] = \varnothing$, then **JEC** is certainly not part of a societal goal, but we could also say that $[\![\mathbf{EC}]\!]$ is an anti-goal. Thus, even though $[\![\mathbf{JEC}]\!] >_J^* [\![\neg\mathbf{JEC}]\!]$, i.e., it is in Jeff's interests for Jeff to exploit children, since $[\![\mathbf{JEC}]\!] \cap [\![\mathbf{EC}]\!] \neq \varnothing$, i.e., $[\![\mathbf{JEC}]\!]$ intersects an anti-goal, it is not in fact an ideal in Jeff's interests.

11.4.4 Group Obligations

So far I have avoided the question of whether group obligations are philosophically coherent. And I will continue to avoid it. The model of obligation applied in this paper is one I have adopted from elsewhere as I have said. It is a model which connects with the 'ought' of game theory, which is usually thought of as an instrumental ought, but in fact is more general than that. It is an axiological ought; an ought based on judgements of value. The obligations groups have in this model follow from comparing the effects groups can have, and what ought to be done is what will have the best effects. But how that is determined is by looking at all of the effects that might occur when combined with the actions of those outside of the group.

If, however, a group doesn't know what the ultimate effects of what they can do are, then they cannot figure out what they ought to do. This uses an objective sense of 'ought' at odds with such an epistemic limitation that I am positing. When game theorists talk of decisions under complete uncertainty, they often posit that the individuals have knowledge of how the effects of their actions interact with others' actions. In real life situations, we barely know what the effects of our own actions will be let alone what will happen when others act. I will be generous, let's suppose that each group can figure out its own capabilities, but they do not know how their actions will influence the effects of others' actions.

Does that make the assumption (c') that $E_B(s, \bar{b}) = \bigcup \{ E_B(s, \bar{b}, \bar{a}) : \bar{a} \in A_{N \setminus B} \}$ problematic? Not necessarily. Whatever the "complement" group does must be contained within $E_B(s, \bar{b})$ when B choose to do $\bar{b}$ according to coalition monotonicity. All (c') means, then, is that every possibility that remains open after B chooses $\bar{b}$ remains possible. I would think that if B thought some state would be excluded by every combination of actions from $A_{N \setminus B}$ with $\bar{b}$, then it really isn't a possible outcome. But that doesn't mean the group B must *know* what specifically effects of those combinations are.

The question becomes whether a group can consider only the effects they alone can have, and then come to the right conclusions about what they ought to do. Can they even figure out what is permitted to do? Again, the answer is: it depends. More precisely, the question I want to ask is as follows: Is it the case that if $\bar{b}$ is such that $\forall \bar{b}'$, $\bar{b}' \preccurlyeq_{s,G,B} \bar{b}$, then could B consult only the effects that its actions have and reach that conclusion? To answer this I will need a lemma.

Lemma 2. *Let I be a set, $\{ X_i \}_{i \in I}$ and $\{ Y_i \}_{i \in I}$ be families of subsets of W indexed by I. Then,*

(1) when I is finite, $X_i \preccurlyeq^v Y_i$ for each $i \in I$ only if $\bigcup_{i \in I} X_i \preccurlyeq^v \bigcup_{i \in I} Y_i$.

(2) If $X_i \preccurlyeq^J Y_i$ for each $i \in I$, then $\bigcup_{i \in I} X_i \preccurlyeq^J \bigcup_{i \in I} Y_i$.

(3) If $X_i \preccurlyeq^S Y_i$ for each $i \in I$, then $\bigcup_{i \in I} X_i \preccurlyeq^S \bigcup_{i \in I} Y_i$.

Proof. Let R be the preference relation that $\preccurlyeq^*$ is based on for this argument. Suppose I is finite and $X_i \preccurlyeq^v Y_i$ for each $i \in I$. By definition of $\preccurlyeq^v$ for each $i \in I$ there is at least one $y_i \in Y_i$ such that $y_i R x$ for all $x \in X_i$. Since I is finite, choose one such y_i from each Y_i (perhaps the axiom of choice is needed when Y_i is infinite). Since $\{y_i\}_{i \in I}$ is finite, it has at least one R-maximal element, call it y^*. Thus, for all $i \in I$, $y^* R y_i$. Suppose $x \in \bigcup_{i \in I} X_i$, then for some $j \in I$, $x \in X_j$ and so $y_j R x$ since $X_j \preccurlyeq^V Y_j$. By transitivity of R, $y^* R x$. Thus, $\bigcup_{i \in I} X_i \preccurlyeq^V \bigcup_{i \in I} Y_i$.

Suppose $X_i \preccurlyeq^J Y_i$ for each $i \in I$. Let $x \in \bigcup_{i \in I} X_i$, then $x \in X_j$ for some $j \in I$. Since $X_j \preccurlyeq^J Y_j$ there is $y \in Y_j$ such that $y R x$. Thus there is $y \in \bigcup_{i \in I} Y_i$ such that $y R x$. Therefore, $\bigcup_{i \in I} X_i \preccurlyeq^J \bigcup_{i \in I} Y_i$.

Suppose $X_i \preccurlyeq^S Y_i$ for each $i \in I$. By lemma 2.2 $\bigcup_{i \in I} X_i \preccurlyeq^J \bigcup_{i \in I} Y_i$. Suppose that $y \in \bigcup_{i \in I} Y_i$, so there is $j \in I$ such that $y \in Y_j$. Since $X_j \preccurlyeq^S Y_j$ there is $x \in X_j$ such that $y R x$. Hence there is $x \in \bigcup_{i \in I} X_i$ such that $y R x$. Therefore, $\bigcup_{i \in I} X_i \preccurlyeq^S \bigcup_{i \in I} Y_i$. □

That lemma will provide the material for the next observation.

Observation 6. *Suppose $\mathfrak{M}$ is a finite choice axiological model that uses J, or S for $\preccurlyeq_G^*$. Then $\mathfrak{M}, s \Vdash O_{G,B}(\bar{b})$ only if $E_B(s, \bar{b}) \in \max(E_B(s))$.*

Proof. Suppose $\mathfrak{M}, s \Vdash O_{G,B}(\bar{b})$. By definition, for all $\bar{b}' \in A_B$ at s, $\bar{b}' \preccurlyeq_{s,G,B} \bar{b}$. So also by definition, for all $\bar{b}' \in A_B$, and all $\bar{a} \in A_{N \setminus B}$,

$$E_N(s, \bar{a}, \bar{b}') = E_N(s, \bar{a}, \bar{b}) \text{ or } E_N(s, \bar{a}, \bar{b}') \preccurlyeq_G^* E_N(s, \bar{a}, \bar{b}).$$

Since $\preccurlyeq_G^*$ on schemes J and S is reflexive, this means we can ignore the first disjunct in the condition above. Given that $E_B(s, \bar{b}') = \cup \{E_N(s, \bar{a}, \bar{b}') : \bar{a} \in A_{N \setminus B}\}$ by (c'), and similarly for $E_B(s, \bar{b})$, by lemma 2.2 & 3, $E_B(s, \bar{b}') \preccurlyeq_G^* E_B(s, \bar{b})$. Because $\bar{b}'$ was arbitrarily chosen, $E_B(s, \bar{b}) \in \max(E_B(s))$. □

What observation 6 tells us is that if an action is obligatory for B, B can find that out simply by comparing the effects of its choices. B needn't consult the combinations of its effects with the actions of $N \setminus B$. However, it needn't be that just because a choice/action is maximal in its effect that it is obligatory for B. That is the converse of that condition isn't the case. The converse does hold, if the scheme H is used.

Observation 7. *Let $\mathfrak{M}$ be an axiological model using H for $\preccurlyeq_G^*$. If $E_B(s, \bar{b})$ is in $\max(E_B(s))$, then $\mathfrak{M}, s \Vdash O_{G,B}(\bar{b})$.*

Proof. Suppose $E_B(s, \bar{b}) \in \max(E_B(s))$. By definition, for all $\bar{b}'$, $E_B(s, \bar{b}') \preccurlyeq_G^H E_B(s, \bar{b})$. For any $\bar{a} \in A_{N \setminus B}$, $E_N(s, \bar{b}', \bar{a}) \subseteq E_B(s, \bar{b}')$ similarly for $E_N(s, \bar{b}, \bar{a})$. From observation 2, it follows that $E_N(s, \bar{b}', \bar{a}) \preccurlyeq_G^H E_B(s, \bar{b}, \bar{a})$. Since $\bar{a}$ was chosen arbitrarily, as was $\bar{b}'$, we have: for all $\bar{b}' \in A_B$, $\bar{b}' \preccurlyeq_{s,G,B} \bar{b}$; i.e., $\mathfrak{M}, s \Vdash O_{G,B}(\bar{b})$. □

What that observation tells us is that using the scheme H, we can always be certain that B's actions correlating to maximal effects will be the obligatory actions. This is the converse of the previous observation.

The forgoing observations also indicate something else. When a group B is obligated to do $\bar{b}$, according to the H scheme, for any subgroup of B, B', the actions $\bar{b}' \subseteq \bar{b}$ will also be obligations of B'. That will also follow from observation 2. Is it possible that what a group ought to do is at odds with what its subgroups ought to do? Consider the following situation.

11.4. TYPE RAISING IN DEONTIC LOGIC

Let $\leq^*_G$ be determined using the scheme S. Let's define the following axiological model $\mathfrak{M}$: Four states: $W = \{1, 2, 3, 4\}$. Three agents $N = \{a, b, c\}$. At state 3, $A_a = \{a_1, a_2\}$, $A_b = \{b_1, b_2\}$, and $A_c = \{c_1\}$. Let's say c's preferences or interests dictate the following relation:

$$1 P_c 2 P_c 3 P_c 4.$$

And finally, let's say

- $E_a(3, a_1) = \{1, 2\}$; $E_a(3, a_2) = \{2, 4\}$
- $E_b(3, b_1) = \{1, 4\}$; $E_b(3, b_2) = \{1, 2\}$
- $E_c(3, c_1) = W$

and the combined action outcomes are given by the following table where c's contribution has no effect:

$E_{\{a,b\}}(3, \ )$	a_1	a_2
b_1	1	4
b_2	2	2

The reader can check that this is a viable specification for the function E. Also, as the reader can check, we have the following relationships between sets of states:

- $\{1\} >^S_c \{2\} >^S_c \{3\} >^S_c \{4\}$, and
- $\{1, 2\} >^S_c \{1, 4\} >^S_c \{2, 4\}$.

One should notice that while according to the type raising scheme J, R_c returns $\{1, 2\} \sim^J_c \{1, 4\}$, the Onto condition in the S scheme means that $\{1, 2\}$ is strictly preferred to $\{1, 4\}$. What that means is that what $N = \{a, b, c\}$ ought to do is what $\{a, b\}$ ought to do (in c's interests), and that is (a_1, b_1). What a ought to do without consulting what influence b might have is a_1, so far so good. But what b ought to do without consulting how a can affect the outcome is b_2. So although $E_b(3, b_2)$ is maximal in $E_b(3)$, it is not the case that $\mathfrak{M}, 3 \Vdash ob_{c,b}(b_2)$. The type raising scheme really matters.

That makes some intuitive sense. If I am making dinner for just myself and my daughter, I make something she will like which I, on my own, can make. But if my partner is also going to help, perhaps there is something that my daughter likes better that we could make together. What has happened in this scenario is that the action b ought to take when acting with a, an action which is maximal in the company of a, isn't even optimal when we ignore a's influence.

So for all schemes except H, if there are obligatory actions, a group can find those by considering just the effects of their own actions, in ignorance of the effects of others. While using H, only considering the effects of their own actions is sufficient for B to determine its obligations. The type raising scheme decides metaethical questions.

As a final technical question I will consider when there are no obligatory actions. If the only optimal actions are those which are permitted, i.e., the actions are not strictly worse than the other actions, none of the schemes is sufficient to determine what those permissions are. The reason is that reasoning to or from $\bar{b} \not\leqslant_{s,G,B} \bar{b}'$ to and from $E_B(s,\bar{b}) \in Opt(E_B(s))$ requires properties that H has and properties that v, J, and S have. That is it requires the type raising scheme to be preserved by subsets and unions of families. When $\bar{b} \not\leqslant_{s,G,B} \bar{b}'$ is true, one of either $\bar{b} \not\leqslant_{s,G,B} \bar{b}'$ or $(\bar{b} \leqslant_{s,G,B} \bar{b}')$ & $(\bar{b}' \leqslant_{s,G,B} \bar{b})$ is the case. In reasoning from $\bar{b} \not\leqslant_{s,G,B} \bar{b}'$ that to the optimality of $E_B(s,\bar{b})$, when $\bar{b} \not\leqslant_{s,G,B} \bar{b}'$ is the case, one needs the property of $\leqslant_G^*$ being preserved by subsets which is only held by H; on the second possibility, one needs the properties in lemma 2. Reasoning from $E_B(s,\bar{b}) \in Opt(E_B(s))$ to $\bar{b} \not\leqslant_{s,G,B} \bar{b}'$ requires that the properties be swapped for the cases. Thus, none of these schemes guarantees that we can reason properly just using the effects of a group's actions. Is there a better scheme? I will leave that question for another day; I have shown elsewhere, that no scheme can have those properties.

11.5 Conclusion

What, I think, these results show is that (1) Schotch's account of social choice pushes one toward consensus in to achieve coherence (i.e., to generate a complete social ordering R_N). The consensus is just of a less overbearing form than dictatorship in the Arrow case. (2) a number of senses of 'ought' can be represented within one framework and compared; particularly, *ought to be* and *ought to do* senses. (3) If it is the case that 'ought to be' does express an axiological ideal in the traditional way, then the notion of type raising can make a difference to how that ideal interacts with what one ought to do. (4) From a limited epistemic position where groups cannot determine what effect the actions of others will have on the effects of their actions, but know what the effects of their own actions will be, the type raising scheme matters in comparing the value of the effects.

I believe conclusions (3) and (4) are important because it shows the possible meta-ethical impact that something as basic as the type raising scheme can have. The choice of type raising scheme is part of the ethical theory and needs to be put into what I call a "medium reflective equilibrium". It isn't narrow reflective equilibrium since in narrow reflective equilibrium one trades off intuitions about the concept "better than" against theoretical understandings of "better than", and those only have to do with how the preference relation is type raised. Those considerations don't seem to tell us which scheme is right. But we don't try to obtain wide reflective equilibrium either since we don't need to leave the realm of value theory, nor is it clear that something that is unrelated to value theory could be affected by something like type raising. What I have shown is that things within value theory, but beyond the concept of "better than" can be affected by the type raising scheme. Perhaps that shouldn't be too surprising; some, e.g., Carr (2015), have already pointed out that the form of type raising is a substantial philosophical/normative assumption which is made by some philosophers and linguists with little argument, e.g., Kratzer (1977, 1991) and Lewis (1973). The concept of "better than" should, indeed, be intimately related to many parts of value theory. But it is not clear how far out variations in the conception of "better than" will be felt. I have shown that it is felt in deontic logic when 'ought to do' interacts with 'ought to be', and when the effects of groups are compared to the effects of their subgroups.

11.6 Bibliography

Abdou, J. and H. Keiding (1991). *Effectivity Functions in Social Choice*, Volume 8 of *Theory and decision lbrary. Series C, Game theory, mathematical programming, and operations research*. Dordrecht: Springer.

Arrow, K. (1951). *Social Choice and Individual Values* (2nd(1963) ed.). New York: Wiley.

Belnap, N., M. Perloff, and M. Xu (2001). *Facing the Future: Agents and Choices in our Indeterminist World*. Oxford: Oxford University Press.

Binmore, K. (2007). *Playing for Real: A text on game theory*. Oxford, UK: Oxford University Press.

Broersen, J. (2014). *On the Reconciliation of Logics of Agency and Logics of Event Types*, Volume 1 of *Outstanding Contributions to Logic*, Chapter 3, pp. 41–60. Dordrecht: Springer.

Carr, J. (2015). Subjective ought. *Ergo, an Open Access Journal of Philosophy 2*.

Goranko, V., W. Jamroga, and P. Turrini (2013). Strategic games and truly playable effectivity functions. *Autonomous Agents and Multi-Agent Systems 26*(2), 288–314.

Horty, J. (2001). *Agency and Deontic Logic*. Oxford, UK: Oxford University Press.

Jennings, R. E. (1974). Utilitarian semantics for deontic logic. *Journal of Philosophical Logic 3*(4), 445–56.

Kratzer, A. (1977). What 'must' and 'can' must and can mean. *Linguistics and Philosophy 1*(3), 337–355.

Kratzer, A. (1991). Modality. In A. von Stechow and D. Wunderlich (Eds.), *Semantics: An International Handbook of Contemporary Research*, pp. 639–50. Berlin: de Gruyter.

Lewis, D. (1969). *Convention: A Philosophical Study*. Harvard University Press.

Lewis, D. K. (1973). *Counterfactuals* (2001 reissue ed.). Malden, Mass: Blackwell Publishing.

Phang, K. (2012, August). Decision under complete uncertainty: Bridging economic and philosophical research. Master's thesis, Dalhousie University, Halifax, Nova Scotia.

Rawls, J. (1971). *A Theory of Justice* (2 ed.). Cambridge, MA: Harvard University Press.

Resnik, M. (1987). *Choices: An Introduction to Decision Theory*. Univ of Minnesota Press.

Schotch, P. K. (2001). Elements of goal theory. In J. Woods and B. Brown (Eds.), *New Studies in Exact Philosophy: Logic, Mathematics and Science*, pp. 245–59. Oxford, UK: Hermes Science.

Schotch, P. K. (2016). Ethics as a formal science. Unpublished Manuscript.

Tamminga, A. (2013). Deontic logic for strategic games. *Erkenntnis 78*, 183–200.

van Fraassen, B. C. (1973). Values and the hart's command. *Journa; of Philosophical Logic 70*(1), 5–19.

www.ingramcontent.com/pod-product-compliance
Lightning Source LLC
Chambersburg PA
CBHW062206080426
42734CB00010B/1808